Dairying with Improved Breeds in Warm Climates

R. E. McDowell

Professor Emeritus of International Animal Science
Cornell University

Visiting Professor of Animal Science
North Carolina State University

Kinnic Publishers

KinnicKinnic Agri-sultants, Inc. Raleigh

First published 1994

by Kinnic Publishers
KinnicKinnic Agri-sultants, Inc.
P. O. Box 37763
Raleigh, NC 27627-7763 U.S.A.

Library of Congress Card Number: 93-81317

ISBN 1-880762-05-6

Printed in the U.S.A. by
Piedmont Litho, Raleigh

Printed on recycled paper

Abbreviations

AI	artificial insemination
AIV	mix of hydrochloric and sulfuric acids
BSC	body condition score
C	cellulose
C_3	three carbon grasses
C_4	four carbon grasses
Ca	calcium
CI	calving interval
CIDR	controlling internal drug release
CL	corpus luteum
CMT	California Mastitis Test
CP	crude protein
CW	cell wall
CWC	cell wall content
DD	days dry
DHI	Dairy Herd Improvement
Dig	digestibility
DIM	days in milk
DM	dry matter
DO	days open
ENE	estimated net energy
F_1	first generation cross of two breeds
FCM	fat corrected milk
GE	gross efficiency
GR	grazing
H	hemicellulose
LH	luteinizing hormone
Max	maximum temperature
MCAL	megacalories
MC/yr	milk credit per year
ME	mature equivalent
Min	minimum temperature
MJ	mega - joules
MOET	multiple ovulation and embryo transfer
NDF	neutral detergent fiber
N-S	north - south latitudes
P	phosphorus
PA	proximate analysis
PRID	progesterone releasing intravaginal device

PTA	predicted transmitting ability
SC	soluble carbohydrates
SCC	somatic cell count
SCCS	somatic cell count score
SV	sire value
TDN	total digestible nutrients
US $	US dollars
WCZ	warm climate zones (N-S 30° latitudes)
WSC	water soluble carbohydrates

Acknowledgements

The content and preparation of this book represent patience and assistance of numerous colleagues and friends. I am most grateful for permission to reproduce published materials and use of photographs.

Special thanks are given to Drs. Ben McDaniel, Steven Washburn and John Wilk of North Carolina State University, Dr. Nelson Philpot, Louisiana Agric. Experimental Station, and Dr. Andres Aluja, Centro De Investigacion, Enseñanza y Extension En Ganaderia Tropical, Martinez De La Torre, Mexico for most helpful inputs on the manuscript.

I am also thankful to herd operators who were willing to evaluate suggestions and who taught me a great deal. Among these are T.J. Trebilcock of Puerto Rico, John Scoffield of Jamaica, Chi-Li Chang of Taiwan and E. Cabello of Mexico. The recommendations on the value of tropical forages arise from the patience and council of Dr. P. J. Van Soest of Cornell University.

Dr. R. C. Gray of KinnicKinnic Agri-sultants, Inc., Raleigh is most graciously thanked for his diligent efforts in organizing the manuscript for printing. The noble efforts in preparation of text and drawings by Mrs. Marian Correll and Ms. Kristy Hill are greatly appreciated.

On Management

"The rule to be observed in this stable at all times, toward the cattle, young and old, is that of patience and kindness. A man's usefulness in a herd ceases at once when he loses his temper and bestows rough usage. Men must be patient. Cattle are not reasoning beings. Remember this is the Home of Mothers. Treat each cow as a Mother should be treated. The giving of milk is a function of Motherhood; rough treatment lessens the flow. That injuries me as well as the cow. Always keep these ideas in mind in dealing with my cattle."

W.D. Hoard
Founder of Hoard's Dairyman, 1885

Preface

The migration of people from rural areas to urban centers of countries in the North-South 30° latitude belt of the world (Warm Climate Zones) continues to rise rapidly. Among the problems in adequacy of food is milk supplies' in these centers. Policy makers are recognizing that the traditional local sources of milk from small farms is inadequate. To support the need for more milk, many countries are importing improved dairy breeds. All countries in Latin America have nucleus herds of dairy breeds, in Asia 70% of the countries and Africa 66%. Over the past two decades, about 500,000 head have been imported mainly from 6 countries in the temperate areas. Currently annual imports are 100,000 to 125,000 head per year which shows there is an expansion. The transfer of high genetic potential cattle for milk production to the "hot climates" has resulted in some successes but unfortunately much less than desired results in many countries. The exporting countries have essentially failed with follow ups on their cattle in the new environments; and nearly all importing countries lack in experienced personnel to guide commercial producers or to train students in dairying at local institutions.

The major objective of this book is to put forth suggestions which can serve both the major needs for decision making for commercial dairy operators and to assist government policy makers in developing strategies on the use of imported dairy breeds. The theses in the discussions arise from years of experience in working with cattle producers in 30 countries of Africa, Asia and Latin America, coupled with evaluations of records of cow performance in 9 countries.

Initially, the writings focus on dispelling two very significant myths about the potential for dairying in the warm climate zones. A wide spread concept is that as long as grasses show green color, their feeding value is good. Secondly, there is the perception that exchanging local types of cattle with new breeds or genotypes with cattle having greater genetic potential for producing milk is all that is required to increase the output of milk from the grazing lands.

The thrust of the second phase is to emphasize the need for consideration of minimum levels of animal biological efficiency as relates to dairy breed cattle. Attempts are made to show that biological and economic efficiency are usually highly related in dairy herd operations. Example of the merits and/or limitations of observations from established herds are given to aid in decision making by herd owners. Additionally, technology from the US and elsewhere are put forth and commented on for application to herd decision making in the warm climate zones, such as suggestions on plans for development of national breeding programs.

I have attempted to be as suggestive as possible and have placed emphasis on commercial dairying to attract and support the investment of private capital in dairying. While taking sole responsibility for any errors of fact or interpretation, I strongly believe that dairying in countries of the warm climate zones constitutes a sector in the national economy which needs inputs to be viable and at the same time achieve levels of efficiency to support food supplies especially for urban centers. A number of drawings are used to illustrate points of discussion along with values in tables to show quantitatively how cattle respond to effects of environmental effects. Also extensive use is made of "Boxes" to provide summations of suggested guidelines for practices and quick reference.

<u>On Development</u>

"The decisive element in development is unleashing the economic potential of the individual , and then more and more individuals so that they collectively make that contribution that eventually becomes so important."

IESC News
(Int'l Exec. Service Corps News)

TABLE OF CONTENTS

TABLE OF CONTENTS (continued)

CHAPTER   **PAGE**

List of Boxes

CHAPTER I

EARLY EXPERIENCES

Background

For the past 100 years and more, breeds of cattle originating in the 40° and above north latitudes of the world have been moved intermittently to countries lying in low latitudes of the world (North 30° to South 30°). During the early years, numbers in the low latitudes were small and held principally by country government institutions. Following a surge in exports from the U.S. and Canada to replenish stocks in the countries of Western Europe, at the close of World War II (1946 to 1950), shipments of dairy breeds were accelerated with country goals of partial or whole replacement of local cattle breeds. Unfortunately success with these introductions was limited because the government institutions were slow to respond to the needs in infrastructure to adequately support these new breeds. There was a general consensus that the major limitation to improvements in production was the low genetic potential of the local cattle, therefore changing to better genotypes was all that was required; a serious oversight of the importance of genotype by environmental interactions. These experiences resulted in little attraction to private investors to participate in growth of milk supplies.

Commencing during the late 1940s, nearly all countries of the central latitudes experienced a high rate of human migration from rural areas to urban centers, with urban growth ranging from 6 to 12% per annum. Currently, one-half or more of the human population in all countries are residing in areas classified as urban, e.g. 87% in Venezuela. In the traditional systems, there was an exchange of milk among households in rural areas, but policy makers came to recognize that it would be virtually impossible to collect and move the low quantities of milk from the small herds (1 to 3 cows) to satisfy the needs of the urban centers. Furthermore, it was recognized that capital investment from the private sector would be needed to obtain higher yielding cattle and equipment for support of a dairy industry. Mexico was probably the fastest mover, but had no milk breeds. Government incentives offered to private investors effectively stimulated a large dairying sector based on Holsteins from Canada and the U.S. There are now 26 countries in the low latitudes which are developing local milk supplies using imported stocks of cattle.

India represents an exception. It has a growth rate in milk supplies of 4 to 6% per annum extending over 20 years based on buffaloes and dairy breed x local breed crossbreds reared mainly in villages. The government has supported the infrastructure for village milk cooperatives under the national program of Operation Flood as an incentive for farmers to invest in both improved breeding (crossbreds) and feeding.

Over the past two decades nearly five million Holsteins have been exchanged in international trade with approximately one-half moving from temperate climate countries to countries in the warm climate zones. All countries in Latin America have nucleus herds of imported dairy breeds. There are at present 70% of all countries in Asia and 66% of the countries in Africa using some imported breeds. Breeds other than Holstein or Friesian were popular imports from 1946 to 1960 but more recently the black and white breeds have predominated. (Hereafter in the discussion, Holstein is used to designate black and white cattle). The importing countries are being encouraged to use semen of progeny tested sires from countries of origin of the cattle, hence attempts are being made to establish and maintain good genetic bases through the transfer of genetics of pedigree selected stock in purchased females and the use of good sires. But, with few exceptions, up to this time, the milk yield of 2,500 to 5,000 kg per lactation and calving intervals of 430 to 485 days in the new locations are below optimum and will not support bank rates of credit.

When dairy operators in the countries of the warm climate zones (WCZ) accept the best germplasm that can be offered for yielding milk, they should be equally astute in the application of supporting technology, mainly feeding. The perplexing aspect is that many of the same farmers have adopted the use of improved varieties of cereal grains, fruits and even cash crops such as tobacco and cotton. The use of the new genetic technology in plants in nearly all instances has been accompanied by following recommended practices on the use of fertilizer, pest control, weed control, and generally land preparation to ensure the technology is successful. Corresponding needs in supporting technology suitable for high producing dairy cattle have not been provided, e.g., there still prevails the perception that Holsteins, Jerseys or other breeds can function effectively with high feeding of low to moderate digestible tropical grasses.

The estimates in elasticity of demand for cereal grains among urban populations in countries of the warm climates currently ranges from 0.20 to 0.30 whereas that for milk is from 0.83 to 1.28. The value of 1.28 shows that demand exceeds supply hence good potential exists for expansion of supplies of milk products. Also it has been clearly demonstrated that as per capita income rises there is a rapid increase in demand for animal products, fruits and vegetables with a corresponding decrease in demand for cereal grains and tubers. There is a positive correlation between growth in cereal grain production and output of animal products, especially milk, poultry and pork. More and more around the world the use of animals as a "sink" or secondary market for the use of grains as feed is inducing farmers to produce grains in greater quantities than usually needed for human foods. In other words, animals are being used to ensure greater stability in grain supplies which serves as a valuable asset to overall agricultural production.

Goals and Achievement

Successful dairying in countries of the WCZ will transpire through investment by the private sector. To attract the required investments will necessitate governments having policies for support. Currently there are over 20 countries in the WCZ which have during the past two or more decades achieved effective programs in dairying using principally introduced breeds. Some of the most significant factors contributing are:

- strong national policy
- adoption of appropriate technology
- expansion of feed resources
- development of feed processing
- effective use of improved animal germplasm
- national milk recording program
- farmer training in use of records in herd management
- medication needs for animal health available locally

After approximately 10 years, it becomes the responsibility of producers to show a comparative advantage from producing milk locally over the cost to consumers from using reconstituted milk made with imported non-fat dry milk powder. Otherwise, all consumers are paying to support available supplies which benefits only the higher income sector of the population. Should high subsidization by governments be required beyond 10 years, no doubt serious political repercussions will arise. However, if the costs of producing milk locally remain above the cost in international trade, there can be alternatives. For example, Mexico provides reconstituted milk to low income areas in Mexico City while the higher income group supports local dairying. India mixes locally produced milk with imported powdered milk and water to produce a "toned" milk containing 1 to 2% fat to extend supplies in urban areas. With the goal of having milk supplies as available as possible and at reasonable cost, the dairying sector must be highly conscious in setting schedules of progress in skills of herd management to achieve and maintain creditability with the public sectors for mutual benefit.

Major emphasis through the remainder of the book will focus on the present and future, not only on technical issues but also social and environmental conservation issues. The production of poultry and pigs in most countries is growing rapidly and almost exclusively through the private sector. Dairying must follow suit in order to become accepted contributors to national goals of self sustainability.

An example of intolerance in the use of a breed like Holstein for dairying is illustrated in Table 1.1. The herd consisted of first and second generation offspring of imports in 1974. Feeding was based on grazing and green chopped grasses. With assumed annual sales of milk of 2,500 kg per cow, the herd was reported as likely profitable. This led to the recommendation that importations could be expanded which proved a false assumption as calving interval was 497 days and days dry 227. On a yearly basis, the returns of milk were only 1,825 kg, a level which could not begin to support the capital invested for importations. The birth weight of the female calves of 30 kg was about 29% less than expected. Growth rate from 0 to 3 months was low and even lower at later ages, hence an overall poor development rate for heifers and low performance later. Of the feed resources, 73 to 75% were used by the cows to provide body maintenance needs. The conclusion is that such a herd could not support credit from the national bank thereby requiring the government to repay the loan from World Bank by taking capital from other parts of the national economy.

TABLE 1.1. Illustration of low biological and economic efficiency for producing milk in a hot, humid area using Holsteins.

Trait	Mean	Desired
305-day milk yield (kg)	2410	6000
Age 1st calving (mo)	34	30
Calving interval (d)	497	410
Days dry	227	130
Conception rate(%)		
1st service	41	50
Calf mortality (0-12 mo)%	18	7
Growth rate		
Birth wt (kg)	30	36
0-3 mo, gain/d (g)	440	500
3-6 mo, gain/d (g)	270	550
6-9 mo, gain/d (g)	230	550

Source: Adapted from Osei, 1991.

The objectives of this discussion are to provide some guidelines for use in furtherance of skills in management and use of technology in the WCZ to make dairying with improved dairy breeds economically viable. The general theme will be that dairy breeds from northern latitudes are generally unsuitable for subsistence farms but can be utilized effectively for at least part of the national milk supplies. Dairy breeds can provide acceptable returns when the environments in which they are placed permit them to perform with reasonable economic and biological efficiency.

CHAPTER II

ENVIRONMENTAL CONSTRAINTS OF WARM CLIMATE ZONES

Geographical

Usually the tropics or tropical areas are perceived as those areas of the world's land mass lying between the Tropics of Cancer and Capricorn, about 21° North and 21° South latitudes. There is no clear demarcation of the "subtropical regions." To have reference points to identify areas where needs in management of cattle are common but differ from the cool and cold areas, the term Warm Climates Zones (WCZ) is used to designate the land masses lying between 30° North and 30° South latitudes. The lower altitudes in the WCZ (< 1,000m elevation) represent physical environments which can have deleterious effects on all domestic animals and especially improved dairy breeds through both direct and indirect pathways.

WCZ in the Americas encompass a small strip of the southern US, all of Central America and the Caribbean Islands and extends south to southern Brazil and small portions of Argentina and Chile. WCZ includes major portions of all countries in Africa. From West to east in Asia, part or all of the countries from Saudi Arabia east to China and Taiwan then south to Indonesia are classed as WCZ. For Oceania, about three-fourths of Australia and the island countries of the South Pacific are within the WCZ (Figure 2.1).

The WCZ are the habitat of more than 50% of the global human population and nearly 60% of the domestic animals and fowl, along with the highest populations of wild animals are exclusive for species like buffalo, camel, alpaca, llama, and capybara. In the WCZ, there is high dependence on animals for both goods and services, thus overall a higher interdependence between humans and animals than in temperate climates, except for pastoral herding above 55° North latitude (McDowell, 1991).

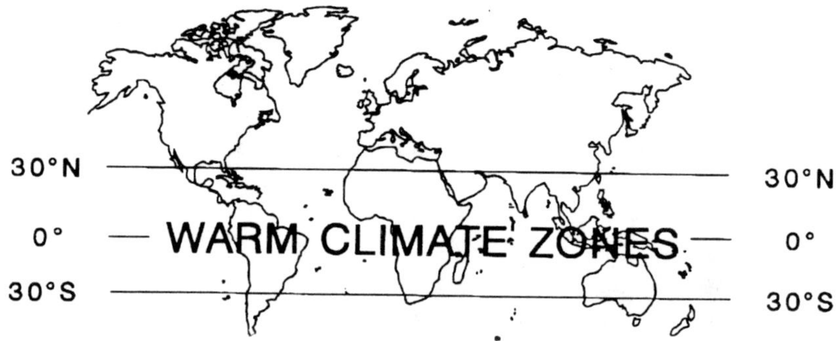

Figure 2.1. **The warm climate zones (WCZ) of the world lie between 30° North and 30° South latitudes (From McDowell, 1972).**

The focus will be principally on areas of < 1,000 m elevation as those areas > 1,000 to 3,000 m have moderate temperatures throughout the year. The management of improved dairy breeds in the higher altitudes can generally make use of the transfer of experiences from the temperate areas 40 to 50° N-S, such as Europe, North America and New Zealand which is not the case for the low elevations in the WCZ.

Perceptions

A general perception is that because a number of domesticated animal species originated in WCZ and large numbers continue to exist, there are likely comparative advantages in feed supplies and climate for livestock production. But as strategies are developed for more output per animal, instead of continuous expansion of animal numbers, some of the perceptions which should be challenged are:

- continuous green color in the grasses means good quality feed is available throughout the year or nearly so

- grazing is always a low cost and a dependable feed resource

- the cost of supplementary feeds should be based on unit cost in weight, not on unit cost of useable nutrients

- all materials on farms should be used as animal feed

- the direct effects of high temperatures on the animals are a greater deterrent than indirect effects, i.e., feed quality

- success in livestock production can be reliably based on a single measure, milk yield

- the genetic potential of local animals is "the" major constraint on productivity

Acceptance of these perceptions will lead to failure of dairying with Holsteins. A major focus will be on working around some of the traditional concepts to create environments for viable economic returns with improved dairy breeds.

Direct Effects

The Holstein breed, as well as most other dairy breeds, originated in western Europe in the region of 50 to 53° N latitude, hence they can be designated as cold tolerant breeds. It is therefore logical to expect that at 35° N where the summers are long there are detectable effects of high temperatures on milk yield and rate of rebreeding and these depressions will show further rise from 30 to 0° latitude. The comfort or thermo-neutral range for the dairy breeds is considered as 6 to 18° C (Figure 2.2). Within this range, there are no measurable fluctuations in their physiological processes and the energy input to output shows good biological efficiency (all body processes will be functioning in their expected ranges). Temperature in the range of -5 to +5° C will often stimulate appetite such that feed

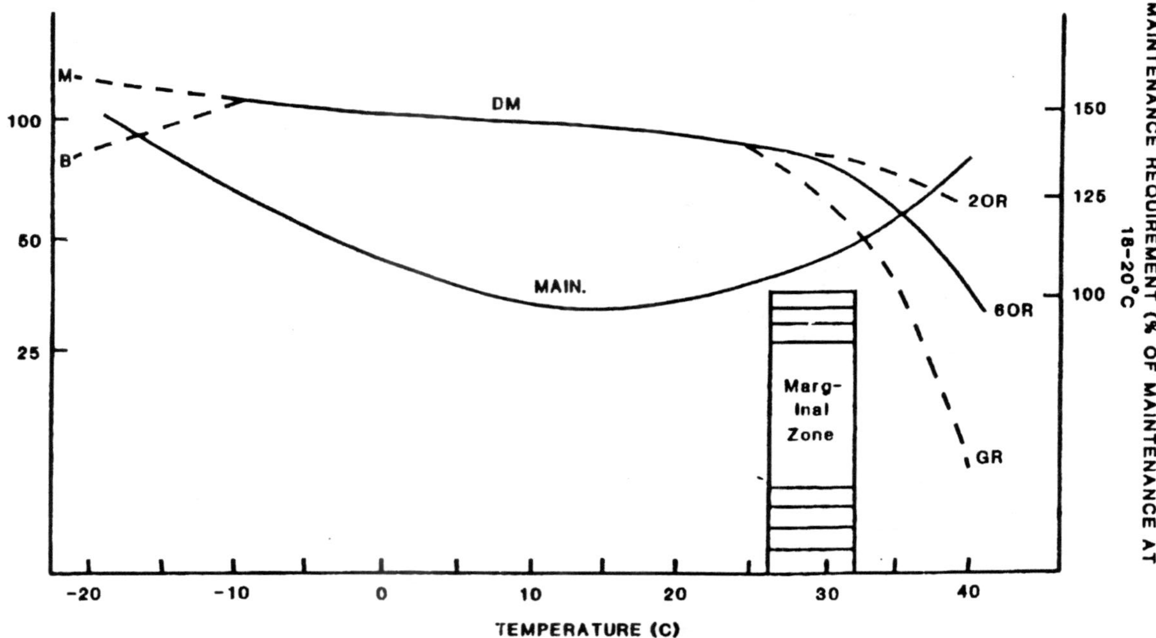

Figure 2.2. Estimated maintenance requirements for 600 kg cow over a temperate range of -15 to +40 °C; percentage change in intake of DM on rations 40% concentrates (60R), 80% concentrates (20R) and grazing only (GR); "M" is needs at -15 to 20 °C with "B" indicating likely intake level due to changes in behavior to conserve body heat; and marginal zone is range for tropics where management is extremely important (Adapted from Nat'l Acad Sci, 1981).

intake will rise more than required to maintain heat balance in the body. The lower temperature range could be the range for best input-output efficiency. Above 27° C, both biological and economic efficiency generally decline (Figure 2.2).

Energy needs for body maintenance are lowest in the range of 6 to 18° C. However voluntary dry matter (DM) intake will remain consistent over a wider range (Figure 2.2); declining at -10° C or below and will be highly depressed at 27° C or above. The extent of the effects of environmental temperature on feed intake or maintenance requirements depends on type and quantity of feed offered (concentrate or forage), level of atmospheric humidity, length of pelage, stage of lactation, appetite and level of milk yield. During the first days of lactation, appetite and rate of feed intake may be more affected by temperature then during later stages.

Important in herd management decisions on feeding are the number of hours of the day the temperature exceeds 27° C and the relative humidity is greater than 80%. When lactating Holsteins are fed free choice a diet of 60 to 65% high quality forages and 35 to 40% concentrate and the temperature is 27° C or higher for 6 hours, feed intake will decline (Figure 2.2). Correspondingly, body maintenance requirements will rise due to increased rate of heat generation in the body from energy needed for greater stroke volume of the heart and increased respiration rate. For example, maintenance needs for the day may rise to 30% higher from noon to about 6 PM than the needs from 2 AM to 10 AM. The net effect on feed intake will depend on the hours < 20° C to allow the cows to restore heat balance in the body.

The level of fiber offered in the feeds also affects the temperature at which feed intake is suppressed (Figure 2.2, 20% R., roughage vs. GR, grazing). The zone indicated as marginal is the temperature range where replacing dietary fiber (GR, grazing) with high unit of energy feeds (20 R) and other management changes, e.g., providing shade or cooling with water spray, can be used to reduce the impact of heat stress. Another important direct effect of heat stress is on the reproduction processes. Effects on reproduction appear both acute as embryonic development is impaired and chronic which affects cyclic status, fertility and gestational blood flow. Exposure for 6 or more hours daily to temperature > 27° C will likely result in a suppression in the output of thyroxine by the thyroid gland. This leads to a lower rate of output of the reproduction control hormones by the pituitary gland thereby a lower presence in the blood. Coupled with this is an increased blood flow to the skin to promote sweating for body cooling. The blood needs water to increase volume. With the onset of heat stress, removal of water from the digestive system and urinary tract rises. After a short period there is some vasoconstriction of blood flow in the central body core. The shifts in blood volume (dilution by water), decreased rate of circulation to the reproductive organs and lower output of reproductive hormones causes in practical terms a "decrease in nutrients" (underfeeding) of those organs which serve to maintain normalcy of functioning in the reproductive system. Management systems to reduce heat load during mid-day can help followed by moving cows into the open after 5 to 6 PM in order to promote maximum rate of conductive heat loss and restoration of body comfort.

Normally, uterine and umbilical blood flow rises with stage of gestation. By 220 days post conception, total blood flow is expected to increase 4.5 fold with umbilical flow about twice that of uterine blood flow. When cows are under heat stress, the uterine and umbilical blood flows become subnormal which leads to decline in growth rate of the fetus. Low levels of nutrition have similar effects.

The importance of using managerial skills on measures to reduce the direct effects of heat for reduced depression in feed efficiency is illustrated in Table 2.1. Gross efficiency remains high irrespective of stage of lactation when exposed to < 20 days of maximum temperature > 27°C. Twenty-one to 40 days exposure to > 27° C depresses efficiency by about 13% in all stages of lactation, but cows exposed 40 to 87 days to > 27° C show medium to marked depression in gross efficiency depending on stage of lactation. This suppression in the efficiency of energy utilization is a further justification of the need for higher levels of managerial skills.

TABLE 2.1. Relation of gross efficiency (GE)[a] of lactating Holsteins when environmental temperature is > 27° C for 0-20, 21-40 or 40-87 days per 100 days of lactation.

| | No. days temperature > 27° C for 6 h | | |
Stage of Lactation (days)	0-20 GE	21-40 GE	40-87 GE
0-100	.85	.74	.62
101-200	.82	.77	.75
201-300	.87	.78	.72

[a] GE = kg milk yield/mcal ENE feed intake.
Source: McDowell et al., 1976.

Limiting Direct Effects

The range of temperature over the entire day can have a significant effect in influencing cow welfare. Average daily temperature for the month of July around San Juan, Puerto Rico, New Iberia, Louisiana (Gulf coast U.S.) and Cairo, Egypt are similar (28° C), Table 2.2. The maximum at San Juan is lowest but there are 16 hours at or above 27° C with only a small decline during the night hours. This means that cows have limited opportunity to restore heat balance. Maximum temperature is higher at New Iberia and Cairo but with fewer hours at 27° C or greater. Nightly temperatures are lower particularly at Cairo along with low humidity. Near Cairo, herds using shelter during the hot part of the day and allowed outside by 6 p.m. maintain feed intake of Holsteins at 90 to 95% of that in temperate areas. However, this standard is more difficult to maintain in warm, humid areas like San Juan.

Seasonal temperature effects on conception rate are demonstrated in Figure 2.3. The mean monthly temperature is quite consistent in Venezuela. The difference between the maximum and minimum temperatures (diurnal variation) is large from October to March but low from April to September. Even though the maximum temperature declines from April to September, the minimum rises because of water vapor buildup in the atmosphere which slows cooling. Thus the cows have considerably less time to restore heat balance resulting in low conception rate.

TABLE 2.2. Hourly temperatures (°C) occurring during July and number of hours ≥ 27° C at three locations.

Hour	San Juan, (Humid) Puerto Rico (18° N)	New Iberia, (Humid) Louisiana (USA)(30° N)	Cairo, (Dry) Egypt (30° N)
00	26	24	22
01	26	24	21
02	26	23	21
03	26	23	20
04	26	22	20
05	25	22	19
06	25	22	22
07	26	23	25
08	27	26	28
09	28	27	29
10	29	28	30
11	29	29	32
12	29	31	33
13	30	32	35
14	30	32	36
15	30	32	38
16	29	32	37
17	29	32	36
18	28	31	35
19	28	30	33
20	27	28	30
21	27	27	28
22	27	26	26
23	27	25	23
AV	28	28	28

The direct effects of heat stress are best countered by providing shades or shelters from about 11 AM to 4 PM (Housing, Chapter VIII). Effective secondary measures are to put cows in open lots with dirt surface soon after 4 PM for the night hours and feed forages after 5 PM and again early AM. The secrets are to restrict heat gains by reduced exposure over mid-day and to allow natural cooling by keeping cows outside post PM milking to near AM milking time.

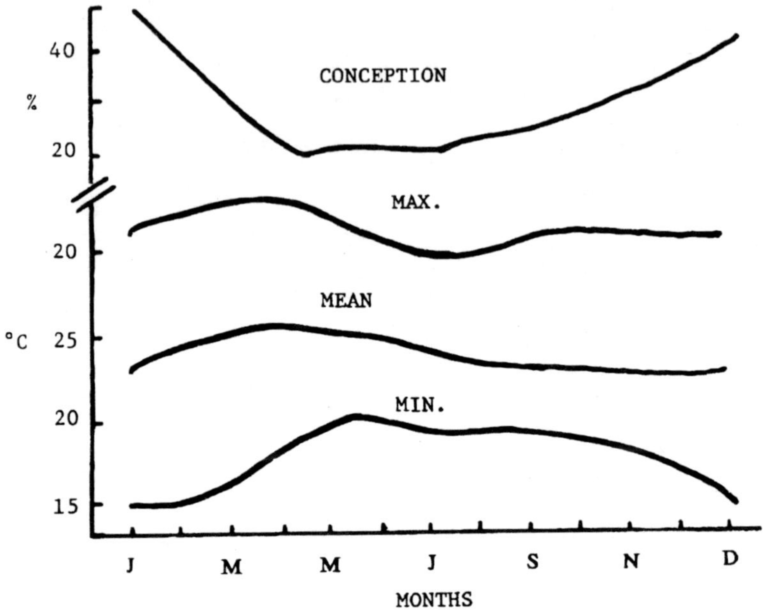

Figure 2.3. Relation of conception rate of lactating cows in relation to variability between monthly maximum and minimum temperatures, Maracay, Venezuela.

Additionally, managing around certain natural diurnal patterns of behavior can be helpful towards modulating heat stress; when do cows prefer to eat, seek shade and wish to lie down (Figure 2.4)? Times of highest feed intake, seeking shelter and lying down are similar for summer and winter which indicate these preferences are ingrained in cattle. Of perhaps greatest significance is the inclination for highest feed intake from 5 AM to about 9 AM and from 5 PM to 7 PM. Working around these peaks, feed intake would likely be highest if the AM milking is finished prior to 6 AM and the PM milking completed before 6 PM. It has also been observed that cows prefer feeding and watering immediately after milking. Overall, shade appears to have its best benefit time from 11 AM to 4 PM. After feedings, a dry surface is needed, preferably dirt, for resting. These patterns in behavior also largely prevail on pasture.

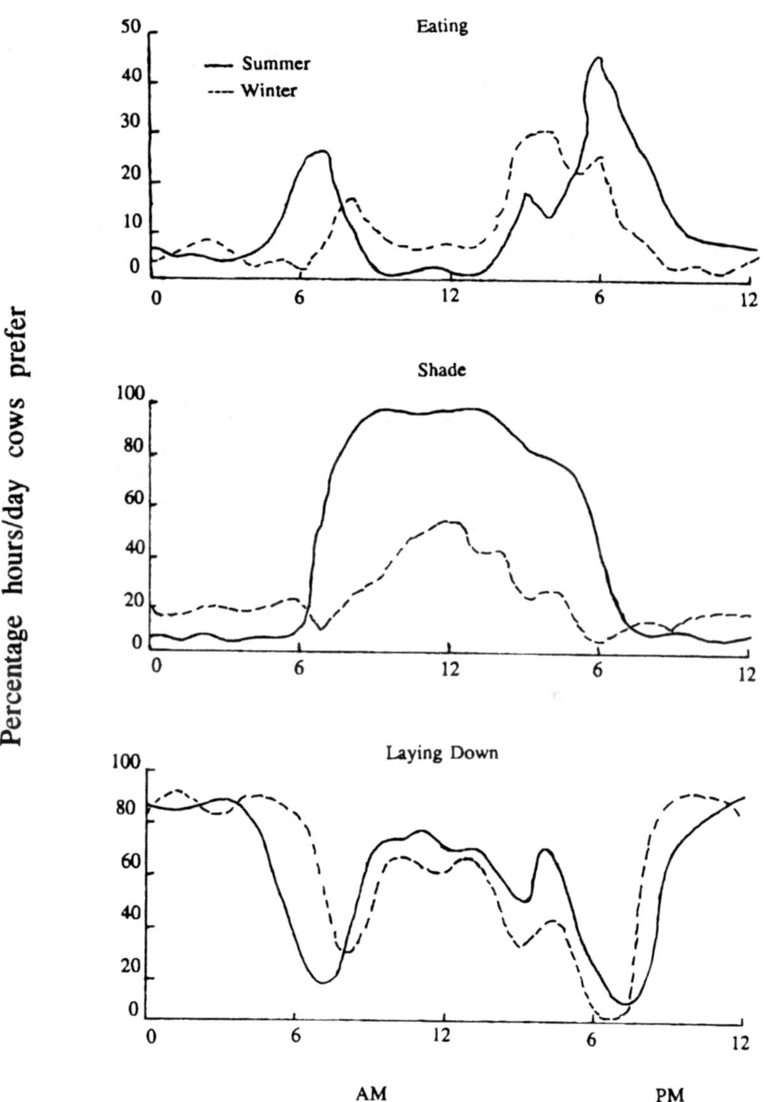

Figure 2.4. Daily diurnal variation in eating, use of shade and lying down in summer and winter.

Indirect Effects

The greatest indirect effect in the WCZ is the influence of temperature on feed quality, especially forages, and to a lesser extent seasonal changes in quantity.

The first stable products of plant photosynthesis of most grasses in the WCZ are four carbon compounds (C_4), while grasses and dicot plants in the 40 to 50° latitudes produce three carbon compounds (C_3).   The difference can be important because C_4 plants are photosynthetically more efficient. C_4 plants tend to exhibit high dry weight accumulations but are frequently low in nutritive value (Van Soest 1982).   However, the generalization that C_4 plants have low feeding value is not universal as maize (corn) and sorghum which are C_4 plants were derived from warm climate ancestors.   These two crops grown in WCZ will nevertheless be 3 to 5 units lower in digestibility than when grown in 35° or higher latitudes.

In the WCZ, the growth of grasses commences after cutting or when rains end a dry period.   Survival of exposure of C_3 grasses to frost demands an accumulation of reserves which provides resistance to cold and increases feeding value.   Such storage is not required in the WCZ resulting in forages lower in quality.   The WCZ environment has led to natural selection for a large proportion of protective structures to avoid predation of insects, diseases and animals (small and large). These protective measures often affect digestibility of the forages.   The general observations are that in comparison to (C_3) grass, digestibility of C_4 grasses by ruminants at 45 days growth (Young, Figure 2.5) averages only 58% vs. 69% for C_3, digestibility of C_4 grasses declines to 45% at the bloom stage but drops to about 38% at 60 days and beyond. During the corresponding period, the crude protein (CP) content may decrease from 15 to 17% to < 5% thereby placing cows consuming near mature C_4 under further nutritional stress. Technology on C_4 grasses has not progressed to the point that large improvements in their quality can be made but some cultivars have been improved slightly through selection. The best approach is to recognize the limitations for feeding of C_4 grasses and complement the constraints by other actions.

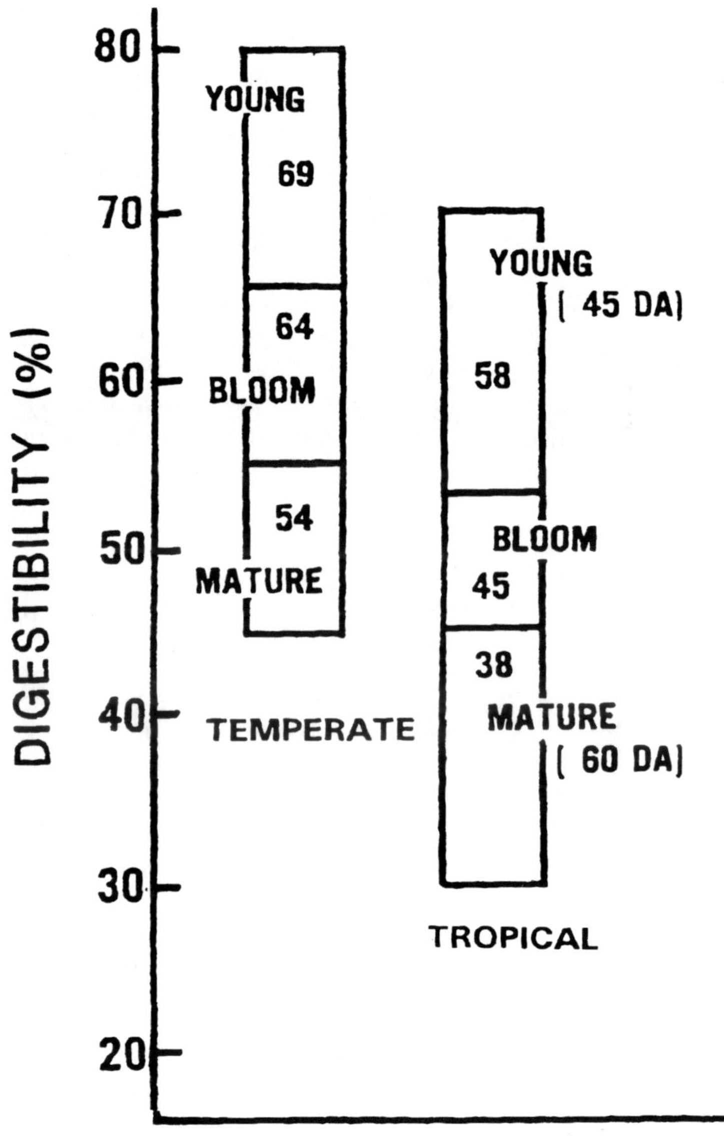

Figure 2.5. Digestibility of temperate (C_3) and tropical (C_4) grasses at various stages of maturity (Source: McDowell, 1972).

As environmental temperature rises, the cell wall content (CWC) of the DM of stems and leaves of both temperate and tropical grasses tends to rise (Figure 2.6), but, the values for tropical grasses are consistently higher. The greater the CWC, the higher the contents of very low or nil digestive fraction (lignin), the medium to low fraction (hemicellulose), and lower content of the readily digestible fraction (cellulose) as illustrated in Table 2.3. Although the temperate grasses show some variability in the four traits, they are on average considerably more useable by cattle than the WCZ grasses. It is noted that the leaves of the three grasses, Cenchrus, Cynodon and Digitaria are lower in CWC than the stems. Except for leaves of Klein grass, the lignin content is higher in the WCZ grasses. The higher hemicellulose (H) and lignin in the cell wall makes it more difficult for rumen microflora to penetrate the cells to release the soluble carbohydrates. The average CWC for the 5 temperate grasses is 58% vs. 68% for the whole plants of the 6 CWZ grasses. The difference will favor daily intake of DM/kg $W^{.75}$ by about 12% due to the longer time the WCZ grasses are in the rumen (Van Soest 1982).

Most soils in the WCZ are highly weathered. Often these soils are low in pH (< 5.3) which, without additions of lime, constrains the types of suitable grasses. Low phosphorus content of grasses in the WCZ is almost a universal problem. Additionally, WCZ pastures may be partially or continuously deficient in calcium, cobalt, sodium, sulfur and zinc. Therefore balancing of rations for minerals requires close attention and includes some minor elements for acceptable biological efficiency of cows.

The environments of the WCZ are also conducive to internal and external parasites as well as certain endemic diseases. Improved dairy breeds can have depression in immune responses from malnutrition through low levels of feeding and imbalances in the feed offered. This increases susceptibility to infectious diseases, the activation of latent viruses, and less than expected response to vaccines. Malnutrition can be a stressor in several ways and affect can impair immunoglobin production particularly when level of protein in the diet is low.

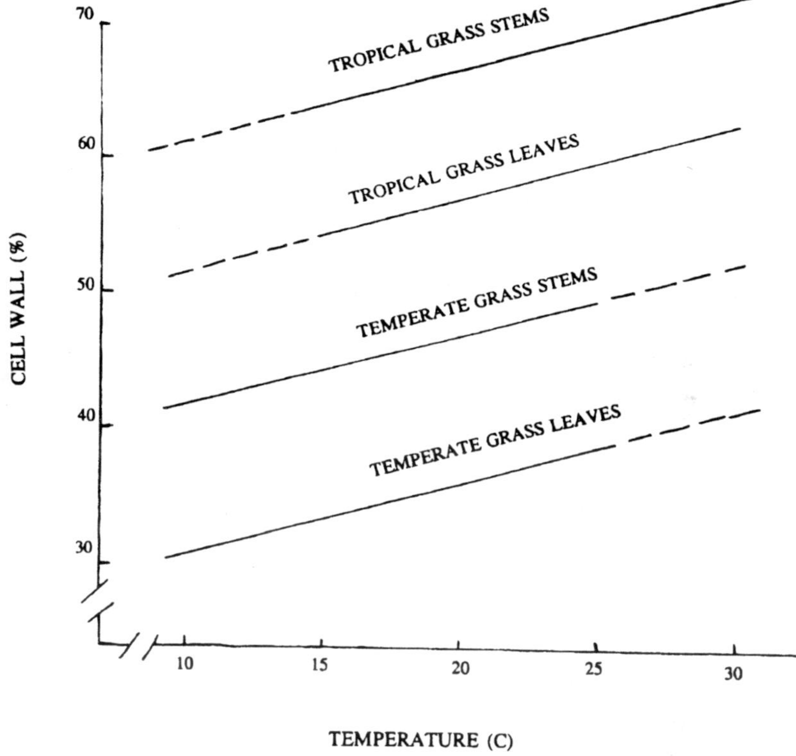

Figure 2.6. Relationship of environmental temperature to cell wall content (% of DM) of temperate and tropical grasses. (Source: VanSoest, 1982). Note: dashed lines are projected beyond actual values from samplings.

The direct effects of heat stress may influence rate of rebreeding but the indirect effects often have greater impact, e.g., low feed energy intake or imbalances in the feeds offered cows will lead to anestrus and abnormal estrous cycles. Delayed conception results in long dry periods. The usual practice when the cows go dry is to feed forages only or perhaps forages with low level of supplement which makes the ration marginal to sub-marginal in CP intake. In dry periods extending 120 days and

TABLE 2.3. Proportion of hemicellulose, cellulose and lignin in the cell wall of some temperate and WCZ grasses.

Genus	Species	Common Name	CW[a]	% of Cell Wall H[b]	C[c]	Lignin
Temperate						
Bromus	*inermis*	brome	64	42	50	7.5
Dactylis	*glomerata*	orchard	55	45	45	7.5
Festuca	K-31	fescue	54	43	46	7.2
Phalaris	*arundinaceu*	reed canary	60	40	51	7.5
Phleum	*pratense*	timothy	57	45	49	4.9
WCZ						
Cenchrus		Klein	66	53	37	7.7
		leaf	65	60	34	5.8
		stem	80	42	45	12.8
Cynodon	*dactylon*	Bermuda	76	51	38	7.9
	nlemfuensis	Star	68	44	45	10.0
		leaf	66	50	40	7.5
		stem	73	40	49	11.1
Digitaria	*decumbens*	Pangola	68	38	48	11.1
		leaf	68	38	50	12.9
		stem	70	37	49	17.2
Panicum	*maximum*	Guinea	66	37	50	8.0
Pennisetum	*purpureum*	Napier	63	34	51	10.4

[a] CW - Cell wall as percent of DM
[b] H - Hemicellulose
[c] C - Cellulose
Source: Adapted from Van Soest, 1982.

longer, the cows may lay on excessive fat from such a diet. This in turn leads to health maladies, such as ketosis, at the following parturition. These "fat cows" often appear as if they will be "world beaters" in production but the vast majority perform well below herd average in milk yield, hence poor feeding can have both short and long term effects.

Cows should show some extra flesh at parturition, condition score of 3.5 to 4.0, to provide energy reserve for use postpartum. Over conditioning can led to a condition commonly called "fat cow syndrome" which will result in a host of metabolic disorders, such as ketosis, displaced abomasum if on high concentrate feeding, retained placenta, mastitis, milk fever, metritis, downer cows and fatty livers. The latter reduces the cow's response to treatment for the other disorders. All these conditions can be costly. Response to treatment can be poor, hence prevention is recommended, along with shorter calving intervals.

When the newborn calf leaves the uterine environment (Figure 2.7), it commences the use of energy to maintain heat balance in its body (homeostasis) to modulate the impact of the ever changing physical environment. In temperate climates, approximately 50% of lifetime feed energy will be utilized to maintain body temperature constant but in the WCZ, about 15% more energy will be required. There is a strong need to execute management practices to minimize stresses. Evidence shows that milk yield is positively related to the proportion of healthy lung tissue especially in WCZ, thus efforts to minimize respiratory problems in the neonatal calf should be of vital concern.

In a practical sense we can readily see the stress imposed by the direct effects of high temperatures. The cows tell us by rapid breathing, salivation, and other forms of behavior. But the "indirect effects" are not often observed or not so visible by behavior until the animal is in serious trouble (Box 2.1). This is the reason considerable discussion has been directed to what is termed the indirect effects arising from the interactions between climates and feeds, diseases and reproduction. As managers, we can help Holstein cows minimize heat stress but our real economic cost with Holsteins is the stresses imposed by the indirect effects.

Numerous other factors could be cited but those included indicate that the transfer of animal germplasm from the temperate areas must be carefully orchestrated. It is repeatedly stated that improved dairy breeds lack tolerance to warm climates. But when fed according to their needs of a balanced ration for body maintenance (4 to 5 kg TDN/day), coupled with at least 10 kg of TDN for production (total/day 14 to 19 kg), the heat tolerance of improved dairy breeds becomes acceptable.

Box 2.1. The Cow: How I adapt to heat stress
                      (Some I show others not)

1.   Increased skin blood flow - Vasodilation
2.   Initiate sweating
4.   Open mouth breathing
5.   Thyroid gland slows body functions through reduction
        of secretion of thyroxine
6.   Change behavior
        •   Reduce eating
        •   Stand in more relaxed posture
        •   Reduce activities including mounting in estrus
        •   Increase water intake
7.   If actions above do not compensate, I must let body
        temperature rise

Figure 2.7. In warm climates, calves will use approximately 15% more of their energy to maintain homeostasis than peers in temperate areas.

CHAPTER III

FEED BUDGET

A feed budget is a plan which involves relating feed supplies to some defined goal for the utilization of a given feed resource, such as pasture or a preconceived goal in animal performance. Possibly the greatest oversight in planning strategies for animal production in the WCZ has been understanding the limits in feed resources. From the mid 1960s through the 1970s there were numerous publications extolling the potential of pastures and rangelands in the WCZ for cattle production. The ensuing recommendations were based largely on yields of DM per hectare and milk yield or animal weight gains per hectare of pastures. The high projected nutritional value was based on estimates from a laboratory method for estimation of energy value termed Proximate Analysis (PA). In recent years it has been shown that estimated nutritive values from PA are reasonably good for C_3 grasses but inappropriate for C_4 grasses. When in the mid 1970s the PA method was replaced by the Neutral Detergent Fiber Method (NDFM), the limits in the utilization of forages of the WCZ began to be recognized. We are now in much better position to sort the merits and limits of WCZ grasses. The following discussion on forages will center on modern technology.

Pastures and Rangelands

Even without any seasonal depression in rainfall, pastures must have species of grasses with distinct differences in growth patterns and time of maturity to sustain yields of dry matter (DM) and quality of grazing throughout the year. Such is quite difficult to achieve in the WCZ. Therefore most pastures consist of grasses with similar growth patterns. All grasses undergo cyclic growth resulting in yields dropping by 50% or more 2 to 3 months during the year even when rainfall and temperature are not constraints. Without a reduction in annual stocking rate to match the environmental limitations, overgrazing, soil erosion, and destruction of the grass stands take place. Frequently for herds in the WCZs, planning of the annual feed budget does not adequately consider the supplementary feeding required to cover identifiable deficiencies in pastures.

Variability in nutrients obtained from grazing is a universal problem in the WCZs (Table 3.1). Estimates of energy intake for the table are based on 300 to 350 kg animals grazing rangelands or pastures with intake of energy expressed as multiples of body maintenance requirements. Those values from 1.1 to 2.3 indicate the proportion of the total energy (10 to 130%) available for productive purposes. All grasslands are good sources of nutrients (> 2.0) for a time (2 to 6 months) but are marginal (low level of nutrients for production) to loss of body weight (< 1.0) for one-half or more of the year.

All locations, except in Puerto Rico, deal with natural grasslands where seasonal rainfall is the most limiting factor to yield of forage. In Puerto Rico (Table 3.1), the pastures consisted of several species of improved grasses fertilized and supplementary irrigation was used when needed. Even with the high levels of inputs, yield and quality continued to be quite variable (1.6 to 2.3 multiples of maintenance or 43% across months). Feeding on the Puerto Rico pastures, a 525 kg lactating Holstein would likely yield 2,800 to 3,200 kg of milk during a lactation period of 280 days depending on month of calving (Yazaman et al., 1982). However, economic efficiency could be low (milk for sale per year) because the calving interval likely would exceed 400 days resulting in a discounted rate of 10 to 15% or an annualized yield of 2,780 to 2,510 kg. Cows receiving supplementary feeding will produce more and have shorter calving intervals thereby higher credit on milk per year.

With a stocking rate of 2.5 cows per ha on the pastures in Puerto Rico, the annual expected return would be about 7,200 kg of milk per ha. This is a good return when compared to < 1,500 kg of milk yield per ha under humid conditions using natural grass pastures, similar to that shown for southern Mexico (Table 3.1, McDowell, 1993).

In much of the WCZ, the predominant family of grasses in the natural grasslands belong to the species *Paspalum* which survive on lower pH soils better than most others. The *Paspalums* are generally shallow-rooted making them susceptible to drought. They are also low growing (< 0.5m) and are low in yields of DM, e.g., Carabao, Dallis, Bahia and Vasey grasses. The

TABLE 3.1. Estimated energy intake by 300 to 350 kg cattle expressed as multiples of maintenance needs from natural rangeland or improved pastures.

Month	Southern Mexico	Improved * Pasture Puerto Rico	Natural Rangelands			
			Northern Nigeria	Central Iran	Central Botswana	Savanna Kenya
J	.8	1.7	.8	.5	2.1	1.4
F	.7	1.6	.7	.5	2.0	1.2
M	.6	2.0	.6	.5	1.7	1.5
A	.6	2.0	.6	1.8	1.4	1.9
M	1.8	2.3	.5	2.4	1.2	2.0
J	2.1	2.3	1.5	2.2	1.0	1.6
J	2.1	2.1	2.3	1.8	.8	.9
A	1.7	2.0	2.2	1.3	.7	.8
S	1.4	1.9	2.0	.7	.6	1.0
O	1.2	1.8	1.5	.6	.6	1.5
N	1.0	1.6	1.2	.6	1.6	2.0
D	.9	1.7	.9	.5	2.1	1.6
Avg.	1.24	1.92	1.23	1.12	1.29	1.45
ADG/ Yr. (kg)	.16-.24	.45-.55	.15-.22	.04-.10	.24-.36	.30-.38
Rainfall (mm)	1000-1200	1400-1700	450-500	200-400	400-500	500-550

*Improved grasses with fertilizer and supplementary irrigation.
Source: Adapted from McDowell, 1988.

Paspalum grasses are usually low responders to applications of fertilizer. Therefore they tend to give best economic returns at low levels of inputs.

Other species, such as Guinea (*Panicum maximum*), the *Cynodons*, e. g. Star grass, and *Digitaria*, e.g., Pangola grass, will respond well to fertilizer and can produce large volumes of DM. However it takes considerable skills to have lactating Holsteins harvest effectively more than 45 to 50% of the DM. These improved species also undergo 2 to 3 months of cyclic changes in productivity each year and if allowed to age beyond 60 days, CP becomes marginal for supporting production. On a sustained herd basis, average lactation milk yield will be in the range of 2,800 to 3,400 kg which would be marginal for good biological efficiency of breeds like Holstein.

Grass-legume forage mixes have merits but there are trade-offs. Application of nitrogen fertilizer to maximize grass production must be restricted to avoid loss of legumes. The legumes may add some soil nitrogen but total yield of DM will be lower thereby supporting a lower stocking rate. At present there are no good models for grass-legume mixes to support commercial dairying in the WCZ. Researchers in Puerto Rico developed recommended grass-legume combinations for pastures but dairy operators with large herds (> 100 cows) chose to use grass stands with nitrogen fertilizer to increase total DM yield per hectare.

There are several important indications. Native grass pastures have limited value as a forage resource for commercial dairy operations. Land requirements are far too high for economic returns. Heavily fertilized grass pastures can be used for feeding lactating cows but their potential on a sustained basis is considerably less than generally projected. Even in the humid lowlands with high rainfall and mountainous areas, marked fluctuation in both quality and quantity of forage are expected. Although well managed, there are variations in DM, CP and digestibility, collectively or individually which limit nutrient intake to less than twice maintenance requirements more than 50% of the time (Table 3.1). Over the short range of a few months, high dependence on pastures may show a favorable feed price to milk price. But under most situations, some form of supplementary feeding can be economically feasible especially when both milk yield and breeding efficiency are included in the evaluation for a total lactation period. (The term TDN is used throughout as a measure of energy instead of MJ as the former is more widely understood among expected readers).

Green Chop

As the amount of productive land in countries of the WCZ lessens, it becomes common to increase the carrying capacity of available land through growing tall, erect, deep rooted perennial grasses. The most popular species is *Pennisetum purpureum* commonly called Elephant, Napier, Merker or King grass. A single hectare of King grass, fertilized and with adequate moisture, can provide for up to 30 head of cattle. These high yielding C_4 grasses have value but are not without problems in supporting milk production.

The first limit in feeding of any of the four *Pennisetums* listed is the level of DM (10 to 20%) while the desired level is in the range of 22 to 25%. Low DM creates several limitations, including transport and handling cost of low DM. The rumen microflora are also inhibited by low DM and high proportion of cell wall content (CWC) in the utilization of nitrogen in the grass to produce microbial protein. Usually the grass is cut at 1.0 to 1.5 m height and chopped for feeding. The fiber content is adequate for needs in total fiber for cows, but with high moisture, the pieces of the grass stalk tend to sink into the ventral sac of the rumen (Figure 3.1) which is below the level most effective to support a rumen mat needed for stimulating the walls of the rumen to initiate regurgitation and further chewing.

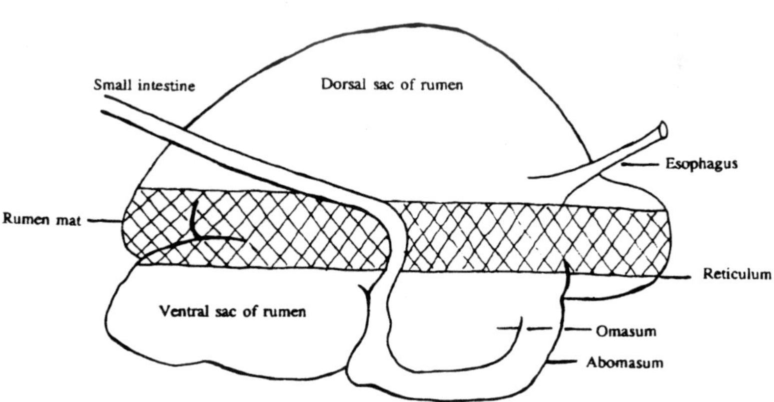

Figure 3.1. A mat of forage pieces 2.5 cm is needed above the ventral sac of the rumen to stimulate the rumen wall to promote regurgitation; mat is determined by NDF content with optimum NDF about 1.1% of body weight (Source: McDowell, 1993).

The "chewing of the cud" serves to enhance the secretion of saliva to serve as a buffer to maintain a normal level of pH in the rumen. The additional maceration of the grass crushes a certain amount of the cell walls to release the cell contents for digestion. When the rumen pH becomes low, an acidic condition arises leading to metabolic disorders due to acidosis and ultimately to health maladies such as ketosis and laminitis. Both ketosis and laminitis are potential sources of stress to cows. The higher the level of concentrate feeding, the higher the risk of rumen disfunction. When the rumen pH falls to between 4.0 and 4.5, the acid destroys the protozoa and bacteria which break down cellulose. Lowering of protozoa and bacteria decreases the efficiency of digestion of forages and the microbial protein which are both important in feed efficiency conversion. The rumen mixture must be stirred to avoid pockets of acids or other products that can reduce bacteria growth. The rhythmic action of muscle contraction and relaxation is an important function in the process. When this slows or stops, there is a problem or soon will be. The conclusion is that maintaining a desirable functioning in the rumen is required in order to maximize fiber digestion.

Buffering agents, such as sodium bicarbonate, can be used to maintain normal rumen pH but this adds cost and does not help the maceration process needed for high cell wall content grasses, e.g., King grass. The feeding of 0.45 kg of sodium bicarbonate per day can serve to reduce acidity, but it is important to be aware that when we feed cows, we are really feeding their rumen's microbes. The rumen microbes provide about 75% of the energy and 60 to 70% of the protein cows need, hence the desire for maximum bacterial growth (McCullough, 1992), (Box 3.1).

Some of the limitations on the use of green chopped King grass in feeding when young or near mature to either lactating crossbred or Holstein cows are illustrated in Table 3.2. It should be noted that maturity of the grass in TDN (young vs. mature) affects level of intake. The DM content in the mature grass is higher, thus some increase in intake of DM occurs (9.9 vs. 10.5 kg for Holsteins) but lower digestibility of the mature grass reduces available TDN. Considering the needs for maintenance,

Box 3.1.   Feeding the Rumen Microbes.

- Rumen temperatures should be maintained at the optimum for bacteria growth and reproduction; another reason to manage for keeping heat stress low.

- Water level high and near constant as the rumen operates best at a moisture level of 80% or above.

- pH or acidity of the rumen must continue in the normal range for the types of bacteria best suited for the fermentation process through adding saliva from cud chewing.

- Rumen mix must be stirred to avoid pockets of acid or other products which reduce growth of bacteria or harm the walls of the rumen.

- Fermentation mixture must be kept supplied with the nutrients required for growth of the microbes.

Source: McCullough, 1992.

only 36% of the total TDN consumed is available for milk production by Holsteins and 41% for crossbreds. This shows that at 52 to 58 kg intake of grass, Holsteins are expected to extract sufficient needs in TDN and CP for maintenance but supplementary feeds will be required to minimize nutritional stress and allow normal functioning of physiological processes.

The tall growing *Pennisetums* are highly differentiated in digestibility from the top to cutting height; growing point and young or mature leaves have a digestion coefficient of 60 to 70%, old, drying leaves 50 to 53%, upper stem 45 to 55% lower stem only 35 to 45% (Figure 3.2). When about 50% of all the chopped grass parts offered only nets about 55% in digestibility, it becomes desirable to offer the cows at least twice

Table 3.2. Estimates of feed intake from young (<2m) and near mature (>2m) of King grass (variety of Napier) and expected performance of 400 kg crossbreds or 525 kg Holsteins.

Measures	Crossbred		Holstein	
	Young[a]	Mature[b]	Young[a]	Mature[b]
Intake				
Total grass (kg)	44.0	40.0	58.0	52.5
Dry matter (kg)	7.4	8.0	9.9	10.5
TDN (kg)	4.1	4.0	5.4	5.2
CP (g)	411.0	320.0	540.0	416.0
Maintenance required				
TDN (kg)	3.1	3.1	3.8	3.8
CP (g)	318.0	318.0	375.0	375.0
Expected (kg)				
Milk from TDN intake	3.1	2.8	5.3	4.7
Milk from CP intake	1.0	1.0	2.0	.5
Multiples of maintenance				
TDN	1.3	1.3	1.4	1.3
CP	1.3	1.2	1.4	1.3
% Yearly TDN to maintenance required	76	78	70	73

[a] Young grass: DM 17% of DM, TDN 55%, CP 10%; intake rate as 11% of body wt.
[b] Mature grass: DM 20%; of DM, TDN 50%, CP 8%; intake rate as 10% of body wt.

the quantity of expected intake. If the cows are forced to eat more of what is offered or have reduced choice, then lower amounts of nutrients per kg of grass consumed results.

For the most effective use of low DM in differentiated chopped grasses, it has been found that it is best to limit the intake of grass and complement it with other feeds. Free choice feeding of chopped grass limits intake of DM. Thus the amount of TDN will be limited, CP level will be quite low and the risk of acidosis in the rumen is increased. The utilization of the nutrients

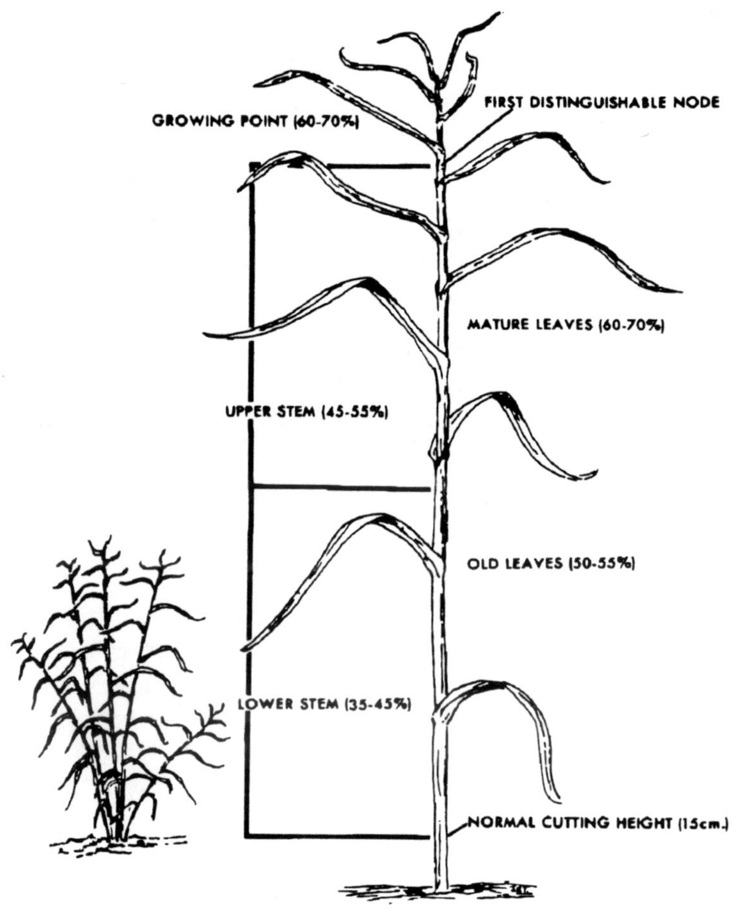

Figure 3.2. Variation in digestive coefficients for portions of the stems and leaves of tropical grasses.

available is restricted and productivity is reduced. A practical approach is to reduce the grass feeding to 25 to 30 kg per day, and supplement with 2 to 3 kg of a dry forage, such as chopped straw or hay, to ensure an adequate rumen mat (Figure 3.1). Additional sources of supplementary TDN and CP which are high in DM (>80%) should also be provided. Wet products like molasses or wet brewers grains will not be suitable to achieve the desired level of DM intake.

Most forages with low leaf to stem ratios and low dry matter content used as green chop have modest to low feed values. These grasses require capital investment in planting, hauling and chopping, therefore skills in terms of cost-benefit analysis should receive attention.

A counter to the discussion on the limits of the grasses in the WCZ will be those who point to the way New Zealand uses grazing alone to support dairying with improved breeds. Those seeing the potential for transfer of the New Zealand technology overlook the sharp contrasts in environmental conditions, such as the cool temperatures in New Zealand throughout the year, which keep plant cell wall content low. Present technology will not permit the transfer of the same grasses; the higher fertility, friability, and pH of the soils; no identified mineral deficiencies, such as phosphorus; and the high use of fertilizer in New Zealand to most of the WCZ. These collectively, coupled with very low differentiation in digestibility of the grasses from ground level to the growing point, all make sharp contrasts to low elevations in the WCZ. However, it is likely that some savings in harvesting and feeding costs can be realized if tropical forages are grazed rather than chopped during appropriate times of the year. This would allow cows to select more digestible portions of plants (Figure 3.2).

Cost of Nutrients from Forages

It is generally found that the cost of 1,000 kg of green cut grasses, from grazing, grass hay or crop residues can be lower than for grains or industrial by-products. The principal is illustrated in Table 3.3. Even after conversion from price by weight to cost per kg of TDN, the cost remains lower (US $0.18 to $0.45 per kg) than most other sources listed. On the other hand, the forages are quite expensive sources of the CP required to balance the ration (US $2.35 for wheat straw, US $4.19 for Napier grass). The price ratios in Table 3.3 tend to over shadow the myth that forages are low in cost in useable nutrients suitable for a balanced diet through noting rank for TDN and CP. For example, crops like cassava could serve more widely as a supplementary energy source but the cost must be low to offset the needed protein supplement. Cassava has 70 kg less CP per

Table 3.3. Estimated cost for 1000 kg and per kg (US$) of TDN and CP for feeds available on farms or in the markets of Pakistan.

Feed	US $/1000 kg	%			Cost ($/kg)[b]			
		DM[a]	TDN	CP	TDN	Rank	CP	Rank
Sunflower meal	337	90	74	45	.51	16	.83	4
Cottonseed cake	265	91	78	44	.37	12	.66	2
Rape seed meal	180	94	76	39	.25	5	.50	1
Rice grain	190	89	79	9	.28	7	2.43	9
Wheat grain	320	88	89	11	.41	14	3.20	14
Barley grain	270	88	84	14	.36	11	2.28	7
Wheat bran	155	89	70	17	.25	5	1.00	5
Wheat straw	75	89	44	4	.19	2	2.35	8
Rice straw	75	89	41	3	.20	3	2.92	10
Rice polishings	240	88	89	9	.31	9	3.18	12
Molasses	96	65	46	4	.20	3	3.90	15
Green cut legume	36	20	66	24	.28	7	.75	3
Sorghum, green cut	54	23	60	8	.40	13	3.14	13
Maize, green cut	54	29	55	6	.33	10	3.17	11
Napier, green cut	47	20	53	6	.45	15	4.19	16
Grass hay (40-50d)	75	92	46	7	.18	1	1.17	6

[a] Estimated % DM as fed.
[b] Cost ($/kg) = estimated useable nutrients converted to $;
1000 kg x % DM = Total DM x % TDN/Cost $ per 1000 kg.
Source: Adapted from McDowell et al., 1990.

ton than maize, therefore must cost $57 less per ton than maize to be comparable in value.

Determining how best to use the forages produced in the WCZ has two implications: 1) there are definite limits for the use of the grasses for feeding improved dairy breeds; and 2) balancing a ration through appropriate supplementation can increase the performance of cows and optimize or decrease cost per unit of production. To provide viable economic returns from Holsteins, plans for feeding must focus on the input cost of useable nutrients of all feedstuffs. It is also critical to ensure that the feeds complement each other to provide balancing of nutrients to maximize use of all feeds. Obviously, feed cost per kg of nutrients is frequently different from cost per kg by weight and both may differ from feed cost per unit of production.

Use of Crop Residues

Limited quantities of crop residues (up to 5 kg per day) can often be used profitably in feeding improved dairy breeds, especially to complement chopped grasses with low DM content (<22%) and wet by-products like brewers grains. For estimates of the amount of nutrients available from crop residues, it is highly desirable to have laboratory analysis of chemical composition because feeding value can be highly variable. For instance, the crop residues from traditional local varieties of cereal grains have digestive coefficients in the range of 42 to 56% while most of the "new improved" varieties are 10 or more units lower (35 to 45%), Van Soest, 1982. The difference arises from selective breeding to increase grain yield and reduce risks of lodging. Stems of the improved varieties are higher in lignin and hemicellulose content, components with either nil or low utility by all ruminants. Dairy operators in Mexico make extensive use of maize stover to complement green cut alfalfa or small grains. They pay up to four times more for the stover of native varieties of maize than the improved varieties because of higher feeding value. There is evidence that selection for increased grain yield in rice, wheat and maize or higher tannin content of sorghum for increasing bird resistance does not mean that the nutritive value of the residues need decline (Reed et al., 1988). However, until plant breeders modify their selection processes, there can be problems in use of crop residues. Much testing has been done on the use of alkali compounds, such as sodium hydroxide or ammoniated compounds, to increase digestibility and raise the nitrogen content. Within the recommended limits of feeding for Holsteins, the cost benefit ratio for either treatment is not usually supportable.

Stored Forages

Greater dependence on the use of forages stored as hay and silage is almost mandatory to promote acceptable levels of performance by cows of improved dairy breeds. However, the quality of the hays or silages need to be such that costs of harvest and storage are adequately covered. As a guide, the cut material (whole plant) should be at least 58% or higher in digestibility and > 9% in CP. Some of the available grasses may fulfill the nutritive value criteria as standing material but by the time the stems are field dried long enough for storage as hay (> 80% DM), leaf losses can become high thereby leaving the end product with a lower feeding value. High levels of management are required to properly store forages and avoid high storage losses.

1. <u>Hay Making</u>

In attempts to increase the value of grasses like Guinea (*Panicum maximum*) or Pangola (*Digitarias decumbens*) for the production of hay, selections of cultivars were based on DM yield. High DM yield led to faster growth rate, earlier maturity and larger stems. These two grasses, which are now common in the higher rainfall areas, can be used to make quality hay (58% TDN, 10% CP). In Taiwan, dairy farm size is small (< 10 ha) but dairy herds large (> 40 head). Other farmers near the dairies are producing Pangola hay which has become highly useful to the dairies. The problem is that the hay producers are cutting for maximum yield (90 to > 100 cm height) but by this stage quality is lowered (50 to 52% TDN) and has also become low in CP (about 5%). However, the demand for hay has been high for use as a supplement with wet by-products, such as distillers grains or pineapple pulp. With the goal of high DM yield, leaf loss in drying is more than desired. Use of a machine to crush the stems at time of cutting would speed drying and reduce leaf loss. On many of the hay farms the grass must be cut by hand due to steep slopes. There are a number of products on the market for spraying on forages cut for hay to reduce respiration in the grasses. The claim of the manufactures is that storage can be undertaken at higher moisture (75% DM) which means the hay needs to lie only 20 to 25 hours instead of 36 to 50 hours. More leaves are saved. Thus far, use of these products on C_4 grasses has been limited because the stems are more lignified than those of the grasses in northern U.S. and Europe where testing of the technology was initiated; nevertheless, further evaluations are needed.

Machinery is now perfected for making a higher moisture hay (often referred to as haylage), about 65% DM. The grass is cut and wilted for about 24 hours, then put into round bales. These bales are wrapped in medium to heavy plastic by machine. Before the final sealing, anhydrous ammonia gas is blown into the bale to help improve the protein content. After sealing, the bales are stored outside and held for feeding almost indefinitely. Preliminary results in Taiwan show the treated hay can be used as the sole forage for Holstein herds averaging over 6,500 kg of milk when adequately supplemented. The high moisture hay could have good utility but cutting lower maturity at (less than 90 cm height) would increase the feeding value.

In brief, the finer stem grasses, like Pangola and Guinea, are acceptable as hay for calves, heifers or cows when cut by 45 days of growth and baled after 24 to 36 hours of field drying. Hay is a good commodity for integrating dairy and crop farms. Hay making appeals to the latter as it provides more frequent income than grain crops. The prediction is that as dairying with improved breeds expands near urban centers because of access to by-products and cost in marketing milk, hay making will rise in more distant land areas as its density can support transport costs. Hay is a needed complement to the silages for good herd performance.

The intake of forages and other fibrous feeds can be increased by grinding or pelleting. A reduction in particle size and collapsing of some cells will increase the density and some in digestibility. However, the rise in intake will vary with the feed. A risk with the use of ground or pelleted forages is reduction in the level of milk fat (Van Soest 1982) and risk of acidosis in the rumen due to lowering of effective neutral detergent fiber (NDF) as will be shown later.

2. Silage Making

Silage making is an emerging technology in the WCZ. Early experiences centered on the use of the tall growing coarse grasses like Napier or Elephant grasses which proved to be poor in value due to oversights of the basic requirements for the ensiling process and high storage losses from low DM content. The grasses were lower in DM (18 to 20%) than desired and low in water soluble carbohydrates (WSC). In temperate areas, moisture is dealt with either by allowing the plant to mature to the point DM rises to 35 to 45% or the plants cut and wilted in the field to the same DM range. Neither can be recommended for grasses common to the WCZ due to loss in quality. If cutting is delayed to raise DM, the digestibility and CP% may be markedly reduced. Cut Napier or King is extremely difficult to pick up with machinery once it is on the ground. Forages low in WSC are prone to secondary fermentation. Products like formic acid and AIV (mix of hydrochloric and sulfuric acids) have been used as substitutes for fermentation of Napier grass but did not work effectively (McCullough 1975). Products like enzymes, bacterial cultures and antioxidants have been tested as means for enhancing fermentation but not as yet to any extent with C_4 grasses.

An adequate level of WSC for fermentation to a stable low pH, lactic acid dominant, well preserved silage is about 3% of the fresh crop weight, but the grasses of the WCZ are generally below 2% in WSC (McCullough 1975). Silage from Napier grass generally undergoes secondary fermentation resulting in high butyric acid content which means the silage is low in palatability, marginal to sub-marginal in CP and not high enough in useable TDN to provide maintenance needs when fed to cattle. A silage high in butyric acid reduces palatability thereby reduces intake of TDN and exaggerates a negative energy balance in early lactation. High butyric silage is also undesirable because the acid is a potential source of ketone bodies which may lead to the condition ketosis, particularly in early lactation. The coarse lignified stems of the *Pennisetum* grasses require a high power input to cut fine enough (0.7 to 1.2 cm) for packing to exclude air. A poor pack leads to spoilage from molds.

Sorghum and maize are both higher in WSC than generally found in grasses. Thus, they have more potential for ensiling and greater proportions of useable nutrients. Maize stalks cut immediately after the harvest of ears for human consumption at the milk stage (roasting ear stage) are better for silage than are many grasses due to higher WSC content. There is a new selection of maize, "tropical corn," becoming widely used for silage in the southeastern U.S. due to its higher tolerance to insects and diseases and longer planting season. The sum is that maize and sorghum selection are expanding use of these crops for forage in the WCZ (Box 3.2).

Satisfactory silages can be produced from wastes and by-products. A successful silage in the Dominican Republic consists of 20% sugarcane tops, 17% molasses, 10% broiler house litter and 51% water. This silage is used as the sole feed for heifers and dry cows and with concentrate supplement is fed to lactating cows. In Mexico, a silage with 40% dry sorghum stover, 20% cage layer waste (wet), 15% molasses and 25% water ferments well and is used as the only feed for heifers and dry cows.

Basically the aim of silage making is to achieve within the ensiled mass sufficient concentration of lactic acid produced by microorganisms in plants to lower pH to a level that will inhibit other forms of bacterial activity thereby preserving the ensiled materials (Box 3.3). Biotechnology has provided commercial products which can be used as an additive during ensiling to expand the concentrations of lactic acid producing organisms. A further advantage of ensiling is that a broad base of materials can be used provided the WSC level is 3% or greater. Ensiling can bring together materials to provide a better end product than when fed separately.

Generally, forages provide the base for profitable dairying. However, the types, quantity and nutrient composition of the forages will depend to a large extent on soil and climate factors. New technology has emerged on forages of the WCZ, but how best to utilize them in feeding high producing cows needs further investigation. Our major concern is that the intake of forages is dependent on the structural volume, therefore cell wall content.

Box 3.2. Tropical Maize for Silage.

- More drought and insect tolerant than temperate varieties.

- Can support milk production similar to temperate maize silages but more concentrate must be fed due to lower grain content.

- Tropical maize needs 110 to 120 days to mature; earlier cut will be wetter and lower in energy thus not support good milk yield; harvest when the starch line is half-way up the kernel.

- At harvest is wetter and even at maturity contains 70 to 75% moisture; seepage not a major problem with mature silages but may be if harvested too early.

- Silage additive may have more value with tropical because is lower in grain and wetter.

- Plant population up to 95,000/ha has supported good milk production and yielded more tonnage and milk/ha.

- Nitrogen application recommended 220 kg per ha.

- Most beneficial role maybe in replacing sorghum in double cropping systems.

Digestibility depends on both cell wall content and the availability of the cell contents as determined by lignification and certain other factors (Table 3.2 and 3.2). Means of maintaining normal pH in the rumen and maximizing the presence of microflora appear as possibilities for more effective use of common grasses grown in the WCZ.

Use of Supplementation

Discussions on grazing, green chop, crop residues and stored forages show that supplementary feeding is required to achieve acceptable degrees of economic and biological efficiency for dairying with breeds like Holstein. Estimates of milk yield in relation to the percent of TDN in the total ration are shown in Table 3.4. A ration consisting of high quality hay and excellent maize silage augmented with a concentrate feed > 80% in TDN can be combined for a > 70% TDN ration. Use of medium quality hay, good maize silage and a concentrate feed of 50% by-products will enable compiling a ration of 60 to 62% TDN. For a ration averaging 55% TDN, a limited quantity of straw can be used (2 to 3kg), but additional forage, 20 to 30 kg, should be a green forage with at least 55% TDN. With the forages 52 to 55% in TDN, additional energy is needed from a concentrate mix at least 62% TDN and 18% CP to achieve 20 kg of daily milk yield.

Table 3.4. Estimated required digestibility of entire ration for expectations in performance.

Expected cow performance	Expected milk yield (kg/d)	Approximate % TDN of total ration
Maximum genetic potential	>35	>70
Intermediate	20-24	60
Medium	12-16	55
Low	3-6	50
Maintenance needs		45
Loss in body weight		<40

Source: McDowell, 1985.

A prevailing perception is that rations high in digestibility are required only by improved breeds. This is untrue as can be seen from response curves in Figure 3.3. A native cow needs at least 4 kg of TDN to achieve a minimum degree of biological efficiency. Using chopped King grass (Table 3.2), a 350 kg native cow would be expected to consume TDN equivalent to maintenance with about 0.6 kg of TDN for production. If the native cow was not stunted from depressed nutrition as a young calf or heifer, on average she will respond through increased performance in almost a linear fashion up to 8 kg of TDN. However an early stunted cow will have a lower response threshold at only 4.5 to 5.0 kg of TDN which is the usual norm for smaller holders. Above 8 kg of TDN the rate of response will slow to a uniform level due to conversion of feed to excess fat (upper limit of genetic potential for conversion of energy to milk production). However, the levelling of responses occurs in all three genotypes.

Box 3.3. The Aim of Silage Making.

Barnett (1954) defines the aim of silage making: "to achieve within the ensiled mass sufficient concentration of lactic acid produced as a result of the presence of microorganisms within the cut crop, to inhibit other forms of bacterial activity and thus preserve the material until used."

Good preservation by fermentation depends on the production of lactic acid to stabilize the silage at a low pH. Such is dependent on an adequate supply of soluble carbohydrates (SC) to produce sufficient fermentation acids to overcome the buffering capacity of the forage. High SC favors acid production. Most grasses produced in the CWZ do not produce good silage due to low SC content, high cell wall content in stems, do not pack well and tend to be low in DM. Easiest to preserve are whole crops, such as maize and sorghum which have modest protein content and large supplies of starch and fermentable carbohydrates.

Under appropriate conditions, silage fermentation will result in preserving over 90% of the harvested energy and protein.

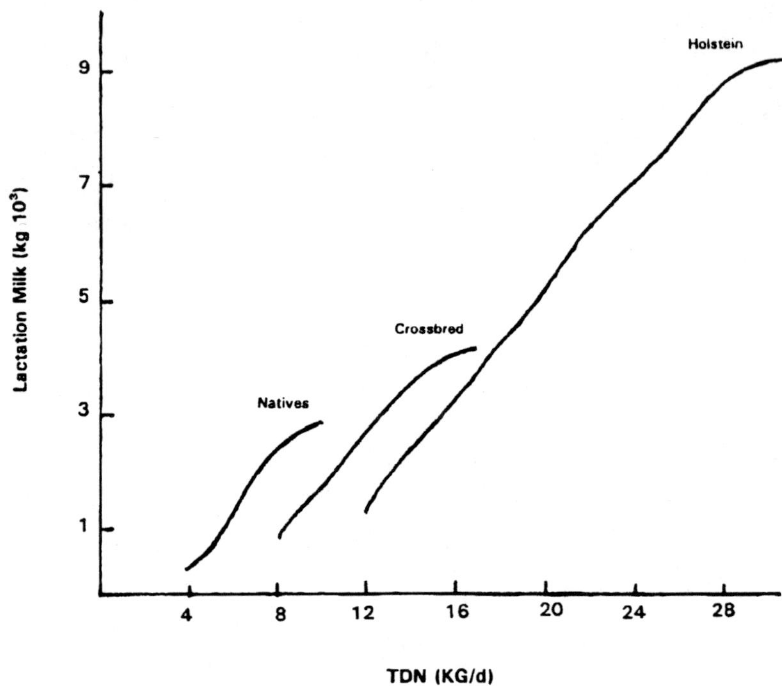

Figure 3.3. Projected response curves from various levels of feeding (TDN) for native, F_1 crosses and Holsteins.
Source: Adapted from McDowell, 1993.

First generation crosses (F_1) from mating Holstein sires to local cows need at least 8 kg of TDN daily and can be expected to respond profitably up to near 15 kg of TDN before efficiency declines (Figure 3.3). A Holstein requires a minimum of 10 kg of TDN, about twice the amount she would be expected to obtain from chopped grass (Table 3.2), for even low performance. Holsteins in many of the temperate areas will respond up to 25 kg of TDN, but in the WCZ their response is expected up to 17 to 20 kg of TDN, depending on temperature stress and digestibility of forages. The differences are attributable to limits on appetite, some increase in maintenance requirements, digestibility of feeds and smaller size cow (525 to 550 kg in WCZ vs. 650 to 700 kg in temperate areas). When a Holstein in the WCZ has an average daily intake of at least 12 kg of TDN throughout a 10 month lactation, efficiency in performance can be good, > 6,000 kg of milk per lactation and a calving interval of 410 to 420 days (McDowell, 1992).

Examples of requirements and merits or limitations of feeding regimes follow for consideration in planning feed budgets for herds. Table 3.5 contains estimates of daily milk yield and requirements of TDN, CP, Ca (Calcium) and P (Phosphorus) by month of lactation needed to produce 6,000 kg of milk per lactation. Daily milk yield in month one is a mean for over 10,000 lactations averaging 6,000 to 6,500 kg per lactation. The rate of decline is on average 12% per month which seems characteristic. (If rate of decline was reduced to 10% there would be a corresponding rise in lactation yield). The estimates of daily requirements include maintenance needs for 550 kg body weight, 3.5% milk fat and from months 2 to 10 for 0.3 kg daily weight gain. During lactation, the cow needs an average of 11.6 kg of TDN per day, ranging from 15.3 kg the first month to 8.4 kg during month 10.

To illustrate the needed supplementation, estimates of intake of TDN, CP, Ca and P from consumption of 24 kg of chopped King grass and 2 kg of wheat straw (source of NDF) are also in Table 3.5. The TDN from grass is equivalent to 88% of maintenance requirements but only 12 to 16% of maintenance for CP, Ca and P. The initial reaction would logically be to offer more grass, but this approach has little validity. Intake of TDN could rise to maintenance needs of 3.4 kg but deficiencies in the other components would increase. To balance out the needs for 6,000 kg of milk, supplements must provide 59% of the TDN, 71% of the CP and over 200% more Ca and P. The example in Table 3.5 supports the desire for dairymen to work toward better quality forages. Preferably these forages would be stored to increase nutritional value and to increase DM intake.

In New Zealand, herds of Holsteins feeding on fertilized grass pastures during the growth period of the grasses average 3,000 to 3,400 kg of milk per lactation. Similar performance can be

Table 3.5. Estimates of daily milk yield and nutrient requirements for 550 kg Holstein for 6,000 kg lactation milk yield.

	Daily	Daily Requirements				From forages[a]				From Supplements[b]			
Mo. of lact.	Milk (kg)	TDN (kg)	CP (kg)	Ca (g)	P (g)	TDN (kg)	CP (kg)	Ca (g)	P (g)	TDN (kg)	CP (kg)	Ca (g)	P (g)
1	34	15.3	3.2	123	78	3.4	.5	3	2	11.9	2.7	120	76
2	30	14.7	3.0	111	71	3.4	.5	3	2	11.3	2.5	108	69
3	27	13.7	2.8	102	65	3.4	.5	3	2	10.3	2.3	99	63
4	24	12.7	2.5	93	60	3.4	.5	3	2	9.3	2.0	90	58
5	21	11.7	2.3	84	54	3.4	.5	3	2	8.3	1.8	81	52
6	19	11.0	2.1	78	51	3.4	.5	3	2	7.6	1.6	75	49
7	16	10.0	1.8	70	45	3.4	.5	3	2	6.6	1.2	67	43
8	14	9.4	1.7	64	42	3.4	.5	3	2	6.0	1.1	61	40
9	12	8.7	1.5	58	38	3.4	.5	3	2	5.3	1.0	55	36
10	11	8.4	1.4	55	36	3.4	.5	3	2	5.0	1.0	52	34
Avg. daily	21	11.6	2.2	84	54	3.4	.5	3	2	7.9	1.7	81	52

[a] Cow offered 30 kg green chopped King grass, 24 kg consumed, 20% DM with 55% digestibility, 9% CP plus 2 kg wheat straw offered and consumed.
[b] Recommended 5 kg of by-product plus concentrate mix.

achieved in a tropical area as shown in Table 3.6. (T_0) for Puerto Rico. Samplings from the pastures in Puerto Rico indicated the nutrients available from pastures would be sufficient for about 4,000 kg production. Tests were conducted to identify the constraints. Supplements of ground maize (T_1) or molasses (T_2) were used to add available energy and to possibly increase the density of rumen microflora to get better utilization of the grass. The level of supplement was kept low (1 kg per 2.0 kg milk > 10 kg yield) with adjustments in supplement on a weekly basis. This meant the cows received supplement for about 8 months. Milk yield increased 14% and rebreeding efficiency improved some even though on average the consumed grass was only about 12% CP and 55% in digestibility. The second stage (T_3, T_4) was to complement the added energy with nitrogen (T_3, 4% urea plus molasses) or protein (concentrate with 20% CP and 72.5% TDN) at the rate of 1 kg per 2.0 kg milk > 10 kg. Milk yield on T_3 and T_4 was 28% above that for T_0 and breeding rate continued within the desired range. The objectives of the four tests on "low supplement" was to increase milk output without reducing the intake of grass, an "additive effect." The efficiency of return (kg of FCM per kg of supplemented TDN) was high, 7.0 to 10.4.

Cows on tests T_5 and T_6 were fed concentrates at the rate of 1:2 kg of daily milk irrespective of level (high supplement) with the differences being in stocking rate (T_5 2.5 cows per ha and T_6 at 5 cows per ha). The T_5 cows produced 46 % more milk than T_0 cows and T_6 cows were 66% higher. On these feeding systems, milk fat percent declined, days open and calving interval rose and efficiency showed a marked drop. The difference in yield between T_5 and T_6 was attributed to overgrowth of pastures with decline in quality at 2.5 cows per ha vs. 5.0 cows. The higher stocking rate kept grass quality higher. In both T_5 and T_6, forage intake declined due to substitution of energy from supplements.

Hoffman et al. (1993) supplemented grazing at the rate of 1 kg of concentrate per 3, 4 or 5 kg of milk yield over 24 weeks of lactation irrespective of yield. Daily milk yield (23 to 25 kg) did not differ significantly but the cows fed 1:3 gained more weight which suggests that rate of supplementation should be related to quality of pasture for best economic returns. Several

Table 3.6. Performance of lactating Holsteins fed grazing alone or grazing with various supplements in Puerto Rico.

Trait	Feeding Systems						
	T_0	T_1	T_2	T_3	T_4	T_5	T_6
Milk yield (ME, kg)[a]	3189	3627	3660	3819	4338	4671	5302
Fat yield (ME, kg)[a]	111	125	129	130	145	155	157
Fat %	3.49	3.43	3.51	3.39	3.35	3.34	2.94
FCM (kg)	2810	3187	3211	3452	3841	3966	4040
Days in milk[b]	268	279	278	277	288	290	299
Efficiency[c]		10.4	9.0	10.6	7.0	2.7	2.3
1st estrus (d)	65	60	53	53	55	61	60
Days open	126	112	112	111	128	134	158
Breeding to conception (d)	45	29	36	41	48	53	72
Calving interval (d)	406	392	392	397	408	414	438
% milk credit/yr	90	93	93	92	89	88	83

[a] ME - mature equivalent basis.
[b] Days yield exceeded 5 kg.
[c] Efficiency - kg fat corrected milk/kg TDN from supplement.

T_0 Graze - grazing alone, 2.5 cows/ha.
T_1 Maize - grazing 2.5 cows/ha plus 1.0 kg ground maize/2.0 kg milk > 10 kg.
T_2 Molasses - grazing 2.5 cows/ha plus 1.0 kg molasses per 2.0 kg milk > 10 kg.
T_3 Urea molasses - grazing 2.5 cows/ha plus 1.0 kg urea (4%) - molasses/2.0 kg milk > 10 kg.
T_4 Low conc. - grazing 2.5 cows/ha plus 1.0 kg concentrate/2.0 kg milk > 10 kg.
T_5 High conc. - grazing 2.5 cows/ha plus 1.0 kg concentrate 20% CP & 72.5% dig./2.0 kg milk irrespective of yield.
T_6 High conc. - grazing 5.0 cows/ha plus 1.0 kg concentrate/2.0 kg milk irrespective of yield.

Source: Adapted from Yazman et al., 1982.

observations can be made: 1) milk production using Holsteins can be economically viable when carefully managed and input costs in land, facilities and labor are kept low, such as in New Zealand, 2) the use of pastures alone will be most profitable in the WCZ when lactation corresponds to the approximate 10 months of highest production from grasses, 3) low level of supplement can quite likely prove profitable but high supplementation will not pay in use of the grazing, and 4) technology on how to make most effective use of the projected nutrients in grasses requires further investigation.

In certain situations, units of metabolizable energy from grazing may have a lower unit cost than supplements from outside the farms. When the cost of protein feeds is high, reliance on grazing may be more practical than using products low in CP, such as molasses or cassava. However, when the genetic potential of cows for milk yield exceeds 3,000 kg, increased yields obtained with supplements moderate to high in CP can be economically viable. Experiences have shown that pelleted protein supplement is the best form for grazing animals (Yazman et al., 1982).

In planning for supplementation, the total ration should be balanced in nutrients suitable for a cow in lactation and producing at an expected level.

An expressed goal of a dairy cooperative in Indonesia is to have their herds averaging 20 kg of milk per day. The recommended feeding per cow per day is given in Table 3.7. None of the herds were being fed to this expectancy. Many were having problems with slow breeding and showed evidence of laminitis in the rear feet. Is the feeding a limitation? Estimates of intake of TDN, CP, Ca and P indicates that in relation to requirements, TDN intake is close to needs (+4%), CP is below (-5%) and so is Ca (-27%) but the level of P is exceptionally high, > 100% of requirements. The shortfall in CP will soon limit intake of total feed and it is obvious that the Ca to P ratio is unacceptable and below the recommended ratio of Ca 1.5 to 1.0 P. Even though total dietary fiber appears adequate, the ration is quite low in NDF, hence the cows are experiencing acidosis. This represents a situation frequently encountered in trying to provide feed based on cost per kg of weight instead of considering the merits and limitations when placed before lactating cows.

Table 3.7. Farm feeding of Holsteins in Indonesia

Feeds	Feed/d (kg)	Estimated Content of feeds			
		TDN (kg)	CP (kg)	Ca (g)	P (g)
Intake					
King grass	40	4.6	.64	48	33
Copra meal	3	2.3	.61	6	18
Wheat bran	2	1.4	.08	2	23
Maize meal	1	.9	.81	5	3
Soya meal	.5	.4	.22	2	29
Rice bran	2	1.4	.26	1	3
Minerals (g)	20				
TOTAL	48.5	11.0	2.62	64	109
Requirements					
Maintenance 550 kg body wt.		4.0	.98	22	16
Milk yield 20 kg/d		6.6	1.78	59	37
TOTAL		10.6	2.76	81	53
Difference (%)					
		+4	-5	-27	+ >100

Source: McDowell, 1989.

A common ration for Holstein herds in Taiwan in 1986-87 is shown in Table 3.8. Expected herd average daily milk yield was 24 kg but actual values were 16 kg of milk, 3.2% fat, 448 day calving interval, lactation length was short (218 days) and days dry 210 days. Based on estimated composition, intakes were sufficient to provide needs in TDN and slightly higher than needs in CP but deficient in Ca and excess in P. The ration was wet as all feeds, except the concentrate, were 20 to 25% in DM. The cows were no doubt experiencing acidic conditions in the rumen due to low NDF. This is supported by the fact that the herds were only averaging 2.4 calvings before culling with the main cause of loss being foot problems. The reason the ration is dated is that when the dairy operators of Taiwan were shown the likely limits in their feeding based on cost per kg of feed, they responded to recommendations. The chopped King grass was reduced to 20 to 24 kg and partially replaced by 3 to 4 kg of Pangola grass hay. Also the brewery and distillers grains were reduced from 12 kg to 8 kg and 0.5% of DM calcium supplement was added to adjust the imbalance in Ca:P ratio. In Taiwan, the Dairy Records Farmer Cooperative is offering computed rations to individual farms to maximize use of locally available feedstuffs without creating hazards for the cows.

The reason for great concern over the form of dietary fiber offered both in Indonesia and Taiwan is to ensure the provision of adequate neutral detergent fiber (NDF) for satisfactory rumen function (Figure 3.1). Dairymen in Taiwan imported baled alfalfa hay from Canada and the U.S. They felt the baled hay was bulky and there was loss of leaves in handling, hence they switched to alfalfa cubes or pellets for lower cost per kg of hay consumed. As shown in Table 3.9, the total NDF in alfalfa hay is excellent in effective fiber for rumen function but the processed forms may be low and non-effective in this respect, especially when fed as a supplement with green cut grasses. The table also shows that several by-product feeds and green cut maize plants can be quite low in effective NDF when compared to total NDF. Grass hays like Pangola and Timothy have a lower effective value than alfalfa but are much better sources of effective NDF than the by-product feeds or processed alfalfa.

Table 3.8. A 1986-87 ration used by dairy herds in Taiwan.

	Feed/d (kg)	DM (kg)	TDN (kg)	CP (kg)	Ca (g)	P (g)
				Estimated content of feeds		
			Intake			
Chopped King grass	30	5.4	2.9	.42	19	16
Brewers grains (20% DM)	6	1.5	1.0	.38	5	8
Distillers grains (20% DM)	6	1.5	1.3	.38	2	11
Soya waste (20% DM)	6	1.5	1.3	.45	6	5
Concentrate	8	6.8	5.8	1.60	24	25
TOTAL	56	16.5	12.3	3.23	56	65

Table 3.9. Percentage NDF[a] (reflecting total fiber) that can be within the desired range but effective utilization is highly variable.

| | NDF% | |
Feed	Total	Effective
Alfalfa hay	40-50	40-50
Alfalfa pelleted	40-50	< 10
Alfalfa cubes	40-50	15-20
Brewers grains	46	10-12
Rice bran	33	10-12
Wheat bran	51	10-12
Whole cotton seed	45-50	18-20
Green cut maize	67	24-30
Wheat Straw	85	50
Pangola grass hay (high quality)	70	55
Timothy grass hay (high quality)	55	52

[a] NDF = fibrous cell wall constituents (lignin cellulose, hemicellulose and fiber bound protein)
From: Composite of sources.

As pointed out earlier, industrial by-products are going to be required as supplements to forage. Lactating cows of all breeds will respond to a daily supplement of 3 to 4 kg of a single product such as cottonseed cake to bring milk yield up to about 5 to 6 kg per day but if offered at 10 or more kg per day, the response in milk yield will not continue to increase in a linear manner due to rising dietary imbalances. Nearly all by-products have an unfavorable Ca to P ratio, high P. An appropriate Ca:P ratio can in theory be made up by adding high levels of calcium. As yet, researchers have not determined the effectiveness of adding high levels of limestone. It could be that much of the calcium would be readily soluble and thereby pass through the cow without very effective contributions.

A major reason for the use of whole gains in concentrate supplements for lactating cows is to ensure maintaining higher calcium intake than from by-products alone or to augment Ca supplements. A low cost concentrate used in commercial herds in Pakistan is in Table 3.10. This mix contains 64.3% TDN, 18.2% CP, 0.57% Ca and 0.45% P. The high proportion of minerals is to provide needed supplementary Ca. Thus far when this concentrate mix is used to complement the feeding of about 24 kg of green cut maize or sorghum, plus 3 kg of wheat or rice straw at the rate of 1 kg of concentrate per 3 kg of milk, yield can be pushed up to nearly 20 kg per day. The same concentrate appears suitable for supplementing heifers and pregnant cows. However, the final verdict is not in as the Ca to P ratio may yet create problems due to low effective use of the supplementary calcium. The central point is that by-products can be employed as supplements but to obtain the desired level of performance for Holsteins, the mineral balances need close attention.

Table 3.10. Example of low cost concentrate mix maximizing by-product use for supplementing green forages and straws.

Feed	% of mix
Cottonseed cake	35
Ground wheat	35
Wheat bran	10
Molasses	15
Urea	1
Minerals	4

Water

Another very important supplement is good quality water. It is the nutrient required in the largest quantity, 75 to 130 kg daily (Ruppel 1993). The quantity needed is influenced by several factors, e.g., level of milk yield (87% water), percentage DM in the feeds, ambient temperature and sodium intake.

According to the National Research Council (NRC 1988), the need in free water for lactating cows can be estimated as follows:

water intake (kg/d) = 15.99 + [(1.58 ± 0.271) x (DM intake kg/d)] + [(10.90 ± 1.57) x (kg of milk/d)] + [(0.05 ± 0.023) x sodium intake g/d)] + [(1.20 ± 0.106) x minimum temperature °C)].

The equation is complex for practical use. The best guideline is to have access to water with individual cows making the choice. Most cows like to drink water after milking. At peak in feed consumption, a peak in water intake follows. Feed provides a second major source of water so on dry feed a cow will consume equivalent of 10% of her body weight. This can be supplied twice per day. To maximize intake from pasture, locate water for lactating cows at or near the milking area; otherwise during the hotter hours of the day, fill from water will be substituted for grazing.

Improving water quality is not likely to induce the same impact as supplement feeds but it will boost production and profit. The cow's digestive system is sensitive to all types of bacterial contamination. The watering facility should be disinfected weekly. This can be done using a solution of ½ cup of chlorine bleach per 5 liters of water. Another rather common occurrence is water can become toxic, e.g., high nitrate levels may result in nitrate poisoning. The conclusion is that more attention to quality and supplies of water will pay. Cooling of the drinking water to serve as relief to heat stress has been suggested but the trade-off is that intake of the cooled water will be lower which in turn influences the efficiency of feed utilization.

Some Indications

Numerous additional citings could be made on supplements but those cited serve to draw attention to this phase in milk production, e.g.,

- supplementation to forage is required for acceptable performance by improved dairy breeds

- long range effects (six months and longer) require evaluation using several traits, such as time of rebreeding and health maladies occurring at the following parturition

- supplementary feeding in early lactation will generally provide economic returns irrespective of the type of cow

- when the limiting nutrient(s) in the forages is identified, much better animal response can be gained with appropriate supplements

- pelleted protein supplement appears the best form for grazing animals in order to obtain some rumen by-pass

- high concentrate supplement for lactating cows will not pay on grazing unless stocking rate is increased and labor reduced

- use of hays and/or silages will frequently make feed supplements more effective

- when we feed cows, we feed their rumens as the rumen microbes provide up to 75% of the energy and 60 to 70% of protein needs, therefore providing adequate NDF to ensure appropriate rumen functions is vital

- good water quality is one of the strongest features of support for good dairy production.

CHAPTER IV

THE BREEDING PROGRAM

The Warm Climate Zones (WCZ) Cow

Holsteins, Friesians or other improved breeds exported to the WCZ have generally represented the average of these breeds in their home country for genetic potential of milk yield. As will be shown later, we do not have evidence at this time to substantiate the long held concept of "tropical degeneration" except perhaps where sires born locally were used in producing subsequent generations and these were randomly chosen rather than being selected by pedigree. However, we can identify "permanent environmental effects" which almost universally inhibit improved breeds from achieving the level of performance comparable to half-sibs in the countries of origin.

Imported bred heifers will be larger at the same age than those born locally. When the import arrives two or more months before expected parturition, the in utero calf will weigh at birth 17 to 20% less than if the same calf was born in the country of origin as a result of stresses from the move. The effects of heat stress during growth and maturation of the heifer born in the new environment will result in a smaller cow. This coupled with coarser feeds results in cows 100 to 125 kg smaller (500 to 550 kg) than expected, e.g., 625 to 675 kg for Holsteins. Body weights from birth to maturity obtained from three areas in the U.S. (Table 4.1, Temperate grades 1, 2, 3 and 4 (low, 1 to high quality environments, grade 4) and from up to 5 countries for each standard of the humid tropics and humid or dry subtropics (Table 4.1, Trop, Humid, Dry) illustrate the variability across environments. The largest females and fastest growers are from selected herds in the northern part of the U.S. (Table 4.1, Temperate grade 4).

Birth weight in the WCZ (Tropics and Subtropics) is usually 36 to 38 kg. However birth weight can be as low as 20 kg for Holstein females because of environmental effects on the dams (McDowell, 1987). The weight range at maturity (72 mo) is 530 to 606 kg (Table 4.1). Using Temperate Grade 3 as the average for Holstein herds in New York State, the WCZ Holsteins are

Table 4.1. Body weights (kg) of Holstein females from birth to maturity in the tropics, subtropics and temperate areas by gradation of management, low (1) to high (4).

| Age (mo) | Trop | Subtropics | | Temperate Grades | | | |
		Humid	Dry	1	2	3	4
Birth	36	36	38	40	41	43	44
1	45	45	48	50	52	54	55
2	62	52	65	68	70	73	74
3	82	83	85	91	94	96	99
4	106	106	111	116	120	123	126
5	130	131	138	142	147	151	155
6	154	155	163	169	174	178	183
7	147	178	187	194	200	204	210
8	154	201	211	218	224	230	236
9	162	221	232	240	247	253	260
10	172	241	253	261	268	275	282
11	183	261	272	282	290	297	304
12	197	285	298	307	316	323	331
13	210	300	308	321	329	329	344
14	224	314	327	337	344	352	360
15	237	328	342	351	360	367	375
16	252	342	356	366	374	382	390
17	267	356	370	380	389	397	405
18	283	370	385	395	404	413	421
19	300	385	400	411	421	429	438
20	316	400	416	428	438	447	456
21	333	415	432	444	455	465	475
22	349	438	452	467	476	492	502
23	365	453	468	484	494	509	521
24	382	458	474	489	500	515	526
25	386	464	479	495	508	520	532
26	386	469	485	501	514	527	539
27	404	476	493	510	522	536	547

Table 4.1. (continued)

Age (mo)	Trop	Subtropics		Temperate Grades			
		Humid	Dry	1	2	3	4
28	411	485	501	518	531	545	557
29	418	493	509	526	539	553	566
30	425	499	516	533	547	561	573
31	433	506	523	540	554	568	581
32	440	512	529	547	561	575	588
33	447	518	535	553	567	582	594
34	454	523	540	558	572	587	600
35	461	528	545	563	577	592	606
36	463	532	550	567	582	597	611
37	470	535	553	572	586	602	615
38	474	539	557	575	590	606	619
39	479	542	560	579	594	609	623
42	483	545	563	582	597	612	626
41	488	548	566	581	600	616	629
42	492	550	568	587	602	618	631
43	497	552	571	589	605	621	634
44	502	554	573	592	607	623	636
45	504	557	575	594	610	625	639
46	506	558	577	596	612	627	641
47	508	560	578	598	613	629	643
48	509	562	581	600	615	631	645
60	527	576	596	616	632	648	662
72	530	585	606	626	643	660	674

Source: McDowell, 1985.

smaller at maturity by 10 to 20%. The weight difference means Holsteins in the WCZ will likely produce less milk but this does not necessarily result in lowered efficiency nor indicate a need to approach selective breeding in a different fashion.

Breeding Plan

Commercial dairies are generally operated by private investors seeking to gain economic returns. A major consideration is to have the herd made-up of cows with genetic potential to respond to the herd feeding and management sufficiently to produce a profit. It is critical the herd owner recognize that in seeking high performance changing the breed or genotype will not produce the desired response unless appropriately fed and managed.

Through the use of frozen semen and fertilized embryos, nearly all herds in the world have the potential of benefiting from selection programs almost any where else in the world (McDaniel and Dentine, 1985). Unlike other challenges in managing a dairy herd, a high profit genetic program may not require increased budget. However, the most profitable sire choices for one herd are not the best for another herd with different management objectives. Irrespective of the objectives for individual herds, the major source of genetic gain comes from selecting sires to produce the herd replacements. Van Vleck (1981) stated that genetic gains in milk yield through natural service sires would not exceed 0.5 to 0.6% per year, but genetic gain from use of selected progeny tested sires from AI could range upwards of 2.0% per year, about a four-fold increase. Much of the subsequent discussion focuses on sire selection and use.

Sire Selection: Local vs. Outside

A point of considerable debate in the WCZ is whether sires should be selected from the local population or the sires chosen from countries where selection is supported by large numbers, record keeping and progeny testing (Wiggins and McDowell, 1993). A report by Sitorous (1982) indicates cows sired by imported semen in Indonesia had higher performance than progeny of local sires. Similar findings come from Mexico (Powell, 1992). The validity of choice rests with the extent of sire origin by environmental interaction effects. The importance of interaction effects has been a major concern of the author. If sire origin effects are large, there could be justification for selection within country or region but if the effects are small or nil the expense may not be acceptable.

Countries with small dairy cattle populations contemplating a local progeny testing program, are reminded that investments are required for record keeping and costs associated with the bulls during the testing as well as the cost of selection. For example, bulls selected for testing in the U.S. have a purchase price of $3,600, feeding and bedding at the AI center averages $3,500, labor $2,800, cost of sampling $1,600 and operating cost of $3,500 for a total of $15,000. The "final cost" of those selected as progeny tested depends on the selection pressure, e.g., 1 in 8 graduated (7 culled) $120,000, 1 in 9 graduated (8 culled) $135,000 and 1 in 10 graduated (9 culled) $150,000, etc. The reaction is that until we can demonstrate conclusively that sire origin effects are large, progeny testing in small populations can not be recommended.

There is the perception that environmental conditions in the WCZ have created changes in phenotypic characteristics which we associate with the survival of local stocks, e.g., the hump and large dewlap of *Bos indicus* or Zebu type cattle and the fat tail or fat rump of certain breeds of sheep found in the middle Eastern countries and in parts of Africa. More careful examination of these features reveals that the characters enumerated were influenced through selective breeding by farmers and pastoral breeders. The hump of zebu enables the use of lower cost harnessing for draft power and the fat tail of sheep are desired to provide a ready or quick source of fat critical to pastoral living (McDowell, 1972). The central point is that as efforts are made to relate features of adaptation to genotype, several factors require consideration, including cost.

Will not selection for milk yield be the best "net measure" of adaptation as milk yield is a composite of actions of numerous genes in individual cows? If the position is negative, then the question becomes what measures of adaptation other than milk yield can be utilized to identify genetic variance among sire progeny; allow required recordings to be collected or determined at relatively low cost on 20 or more progeny per sire; and be a trait with a heritability of 20 to 30%? High body temperature, for example, can be related to the appetite of cows, their milk yield and rate of conception, but heritability is low, < 10%. Zebu cattle in the WCZ may have thinner skin than European breeds but the data are confounded with environmental effects (cold, thicker hide; hot, thinner). Zebu cattle do have denser and shorter hairs in their coat, two or more times greater number of hair follicles and about 50% of the length found in European breeds per area of body surface. With a sweat gland and a sebaceous gland associated with each hair follicle, Zebus also have more total glands that can aid in promoting heat loss (McDowell, 1972). The denser hair coat can increase resistance to ticks due to more difficult penetration of the skin and shorter hairs can aid in cooling the body, but we currently have no within breed estimates of variance for these traits to add to a selection index. More recently it has been shown that probably the most significant difference between Zebu and improved dairy breeds are attributable to differences in feeding behavior, particularly on grazing as pointed out in Chapter II. As far as can be determined at this time, grazing behavior is genetically associated as the level of selective grazing is correlated with the proportion of Zebu breeding in crossbreds.

There are other factors important to survival in both the WCZ and temperate areas which have been identified as having genetic relations but heritabilities are low, $\leq 10\%$. For instance, progeny of certain sires have more problems with bad feet or poor udders and higher somatic cell counts in the milk. Some animal breeders have considered dystocia as being genetically related but the general consensus is that it is a "mating problem" since it occurs almost exclusively in heifers. As technology advances, progeny testing may include sire values for these traits which can be used to considerable advantage in sire selection. For example, Holstein sires in the U.S. are now given a predictive value for high or low somatic cell counts among their daughters. Certain features of body conformation (type score) have genetic linkages and can serve to predict survivability to an overall extent, especially for udder characteristics. The recommended process in sire selection is to focus on transmitting ability for high milk yield modulated or indexed by the "risk cost" of achievement. If a sire has potential for high milk yield, will his progeny averages be lower in percentages of milk fat, protein or solids-not-fat which would affect economic factors in total returns from milk sales? Does this sire tend to produce large calves (dystocia) when bred to heifers and do his progeny tend to have shorter herd life because of bad feet, poor udders or mastitis as estimated from somatic cell counts? Information on these supportive traits are important, thus the decision becomes the choice of a sire for high milk yield with above average risk in other traits versus choosing a sire with perhaps lower transmitting ability for milk yield, but having higher merit in certain of the supportive traits. Most artificial insemination organizations (AI) with extensive recording programs, such as the U.S., can provide ratings on a number of the supportive traits. Due to the low heritability for these traits, large numbers are needed to establish rankings for sires with an acceptable level of confidence. With the exceptions of Colombia, Mexico, Puerto Rico, Taiwan and Zimbabwe we are not yet at the stage of sufficient numbers in countries of the WCZ for reliable assessments of sires even for milk yield.

Although sire indexes have not as yet been developed for the use of feed energy, e.g., Bauman et al. (1985) conducted a review which showed that variation among cows was high only for nutrient partitioning and intake (Table 4.2). This was the only efficiency component projected for change through both selection and management. Management practices can be a major factor in the efficiency of digestion and nutrient absorption. Even though we do not have direct sire values for portioning feed nutrients, there is general consensus that sires whose progeny are high in milk yield do transmit best in this respect.

Although heritabilities for disease traits are low, possible genetic variation for disease incidences can be economically important thereby warranting inclusion in breeding programs. However, recordings in the US through the National Health Monitoring System should in time be useful in creating helpful guidelines. At best, genetic improvement in resistance is a slow, long-term process (Shook, 1989). It does not reduce the requirement for good sanitation and management. Where vaccines are available, genetic improvement for one or more diseases will not likely be cost effective. The reason for largely modest efforts on selection for resistance to diseases is that research until now shows resistance mechanisms in cattle suggest a large number of gene loci are involved.

The low heritability of disease traits indicates little opportunity for genetic change through selection on individual performance. However, progeny testing for disease traits, is attainable with reasonable accuracy particularly for comparisons on susceptibility. Sire progeny differences in somatic cell count score and incidence of clinical mastitis are presently a major thrust by researchers in animal breeding in the U.S. and Western Europe. Preliminary predicted differences among sires for susceptibility measures are emerging, hence including sire ratings on susceptibility can be included the final selection of sires.

Sire Origin Effects

Evidence with Holsteins show that differences between the proportionality of variance components for milk yield between temperature climate areas and the WCZ is likely low. In practical terms, this means that the progeny of sires will rank essentially the same for milk yield under both conditions. The percentage of variance in milk and fat yields attributed to sire (5 to 9%) and cow (30 to 40%) in WCZ countries are within the range of values for temperate areas (Abubakar, 1987a,b, Camoens et al., 1976, McDowell et al., 1976, Romero et al., 1989, Stanton et al., 1991).

Because of the large herd-year variation in average milk yield (20 to 40%) generally occurring in countries of the WCZ, more progeny per sire are needed to obtain acceptable confidence in calculated values for sires. In other words estimated differences among sires are not highly reliable. The shift in ranking based on first versus second lactation records of the top and lowest 5 sires among 71 Holstein sires with 5 or more progeny in Venezuela (Table 4.3) illustrates the strong effects the environment can have on estimates of sire breeding value when numbers of progeny are low. Progeny of the high sires for first lactation calved in a good season but calved in a poor season for second lactation. It is noted that the top sire in lactation one dropped to number 66 ranking in lactation two and number 71 sire rose to number one in second lactation. Which is the best sire?

Table 4.2. Sources of variation and prospects for improvement of productive efficiency in dairy cows.

Efficiency component	Among Cow Variation	Sources for improvement	
		Selection	Management
Digestion and nutrient absorption	Low	Minor	Major
Maintenance	Low	Minor	Minor
Metabolizable energy use for milk synthesis	Low	Minor	Minor
Nutrient partitioning and intake	High	Major	Major

Source: Adapted from Bauman et al., 1985.

Table 4.3. Sire values for the top 5 and lowest 5 ranking among 71 Holstein sires used in Venezuela based on first lactation milk yield and shift based on only second lactations.

Sire Code	1st lactation		2nd lactation	
	SV[a]	Rank	SV[a]	Rank
8259	349	1	-97	66
3311	247	2	-6	39
7940	208	3	87	7
4536	207	4	-44	57
2994	201	5	-42	56
6406	-250	67	49	15
9339	-260	68	62	11
0083	-343	69	73	10
2880	-418	70	20	29
2596	-487	71	193	1

[a] SV - predicted sire values for milk yield.
Source: Adapted from Schneeberger, 1982.

Henderson (1971) demonstrated that even in temperate areas the correlation between sire proofs and true breeding value increased rapidly with increasing numbers of daughters up to 50, then levelled off. His recommended minimum was 20 daughters per sire in 10 or more herds in temperate areas but with large year differences, 40 to 50 daughters per sire in several herds may be needed for the equivalency in confidence of breeding values compared to more stable environments. Sire evaluations are important as up to 76% of the total genetic progress comes from sire selection due to their contribution from use as sires of cows and as sires of sires (Van Vleck, 1977).

The genetic correlation in performance of progeny by U.S. proven AI Holstein sires in the U.S. and Mexico was high (0.86) but strongly related to number of progeny (five daughters 0.65; 10, 0.75; and 20, 0.86) (McDowell et al., 1976). The authors concluded that the magnitude of sire by origin effects should be small.

For over two decades, breeders in Mexico have selected bulls from local herds for progeny testing. In 1988, there were 628 sires having $\geq$ 10 progeny with milk records in herds of Mexico. Of these, 116 originated in Canada, 339 came from the U.S. and 173 were pedigree selected from herds in Mexico (Table 4.4). The percentages of sires with plus (+) value in PTA milk were 10 Canadian, 75 U.S. and 15 Mexico. Average PTA milk values also favored U.S. sires by 306 kg versus 128 kg for Canadian and 153 kg for Mexican sires. The proportion of cows not completing a first lactation was lowest for U.S. sires (8.7%). Had the progeny of sires from herds in Mexico showed lower losses during first lactation, they could have been deemed more adaptable. The lower number of sires from Canada with +PTA milk was because of quite high emphasis given to type score in selection. The ranking of the three sire sources in 1988 has prevailed since the first summary in 1978. Mexico does not offer advantages to select local sires, except perhaps in cost of semen.

Using a single herd as the origin of sires for a population can be important due to accumulation of inbreeding effects. As an example, 44 sires of the Jamaica Hope breed coming from a

Table 4.4. Performance of progeny of Holstein sires from Canada, U.S. and Mexico in herds of Mexico.

	Canada	U.S.	Mexico
Number			
Sires with $\geq$ 10 dtrs	116	339	173
Dtrs with records	9,697	26,455	8,163
Sire, plus PTA milk	42	304	59
Dtrs per sire	82	78	47
Percentages			
Sires plus PTA milk	10.3	75.1	14.6
Total sires used (+)	6.7	48.4	27.5
Incomplete 1st lactations	9.7	8.7	11.5
Av. PTA milk (kg) for + sires	128	306	153

Source: Powell, 1988.

single government owned herd in Jamaica were used in 28 commercial herds of Jamaica during the period 1969 to 1983. There was an average of 23 kg per year downward genetic trend in the herds (-345 kg) attributed to a corresponding rise in the inbreeding coefficients of the sires stemming from a relatively small herd of 180 to 200 cows (Abubakar, 1983). The decline in genetic potential is as expected as the level of inbreeding rises. Data have not yet become available to conduct a similar study on Holsteins in any country of the WCZ. In Puerto Rico, using AI sires from the U.S. and some bulls from local herds there was a positive genetic gain of 7 kg of milk per year in herds averaging 4,860 kg of milk per lactation favoring progeny of U.S. sires (Romero et al., 1989).

Profitability in Sire Selection

Given that it would be desirable to import semen of progeny tested sires, there is the question of whether the practice will be profitable? Evaluations made by Blake et al. (1988) and Holman et al. (1990a) will serve as case studies.

The objectives were to quantify the economic returns from investing in U.S. Holstein AI sires applying the net present value (PV$) sire model for Colombia, Mexico and Venezuela to determine the profitability of U.S. sires. Returns on investment in U.S. sires over time in the herds and lifetime returns were calculated as PV$. As shown for Mexico in Table 4.5, PV$ was determined for two selection policies (SP) using relative ratios of importance on PD milk (1:0 and 4:1) emphasis for milk and type score (PDT), 30 and 50% conception rates to first service (CR), three real interest rates (RIR), 3,7 and 10% and herd averages for milk in Mexico and the U.S. A summary for 198 U.S. sires available for purchase of semen in Mexico in 1987 was made. Table 4.5 contains the averages for PV$, semen cost (average $18.00 in Mexico) and PD type, (PDT) for SP 4:1. Averages for these same sires in the U.S. are in parenthesis for comparison. Using a RIR of 7%, which was the average profitability of U.S. sires in Mexico, returns for sires were less to Mexican breeders than in the U.S. with 50% CR and 3% RIR despite crediting progeny in Mexico for a lower genetic base. However, most bulls (60-90%) were profitable (PV$ > 0) for Mexico using the SP, CR and RIR considered. Profitability of U.S. semen diminished with decreasing CR and rising RIR in Mexico. The indication is that Holstein breeders have abundant opportunity to earn good returns from investment in U.S. semen. The level of return, however depends on level of CR and RIR. The 20 most profitable bulls in Mexico at 30 and 50% CR and RIR of 7% averaged 98 PV$ and 122 PV$ for the 1:0 SP and 109 PV$ and 134 PV$ for the 4:1 SP. A similar calculation for U.S. AI sires in Colombia and Venezuela revealed lower profit margin due to less milk per lactation (< 4,000 kg), longer calving intervals and shorter herd life. This study draws further focus on the need for good breeding efficiency to support profitable dairying.

A possible further advantage in the use of imported semen is its quality. It is well known that semen quality of dairy breed bulls declines under heat stress with the greatest effect on the first stage of spermatogenesis which has a delayed effect on fertility. Delayed sexual maturity, testicular hypoplasia and lower libido appear more common among bulls reared in WCZ, therefore bulls should be screened carefully (Galina and Arthur, 1991). Air conditioned housing of bulls will lessen some of these effects but increase cost.

Table 4.5. Mean net present value (PV$), percent of bulls with PV$ >0, semen prices for all bulls and those with positive PV$ and predicted differences for dollars (PD$) and type score (PDT) for 198 Holstein sires in Mexico for 1:0 and 4:0 selection policies (SP), 30 and 50% first service conception rates (CR) and 3%, 7% and 10% real interest rates (RIR).

Combination			PV$[a]		Semen Prices			
SP[b]	CR%	RIR%[c]	All	> 0%	All ($ Unit)	PV$ > 0 ($ Unit)	PD$[d]	PDT[e]
1:0	30	3	1	82	18	9.13	108	
		7	-30	76	18	8.59	108	
		10	-47	69	18	8.11	108	
	50	3	70(54)	92(93)	18(13)	10.82	108(102)	
		7	37	88	18	10.14	108	
		10	18	84	18	9.44	108	
4:1	30	3	30	86	18	9.42	106	.97
		7	7	82	18	8.81	106	.97
		10	-29	79	18	8.23	106	.97
	50	3	101(68)	97(95)	18(13)	12.48	106(102)	.97
		7	60	94	18	11.51	106	.97
		10	37	91	18	10.04	106	.97

[a] PV$, milk price of $.181/kg of milk with 3% fat and differential of $.061/0.1% change in fat.
[b] 1:0, emphasis only on selection for milk; 4:1 milk to type score (PDT).
[c] RIR, real interest (discount) rate 3, 7 and 10%.
[d] PD$, prices of milk and components on national averages used in the USDA sire summary.
[e] 191 bulls with PDT type.

Source: Adapted from Holman et al., 1990.

Progeny Tested vs. Pedigree Selection

Another factor which can affect cost of breeding is the degree of emphasis in sire selection. Selective breeding programs with Holsteins and Jerseys were initiated in North Carolina herds (35° North latitude) in 1972. Progeny tested Holstein sires were chosen from all active AI sires in the U.S. with repeatabilities for PTA milk of $\geq$ 60%; half were chosen on PTA 3.7% fat-corrected milk (FCM) to sire the Milk line. The other half was chosen on an index which gave emphasis to body conformation scores for udder, feet and leg characteristics weighted at one-third the value of PTA 3.7% FCM. This group produced the

Index line. A Young sire or pedigree line was developed with the main selection on pedigree merit for 3.7% FCM computed as 0.5 times (sire's PTA + dam's cow index). Since 1972, each year from 3 to 10 proven sires in each of the two lines and from 6 to 12 young bulls were used.

Comparisons of the lines in the first four generations for various traits in first lactation are in Table 4.6 and the averages for breeding values computed for all lactations by line and birth year are in Figure 4.1. Genetic gain in yield of 3.7% FCM was about 100 kg per annum or near 1.3% of the mean which is near theoretical progress. Thus progress in milk yield through selection has been high with no significant changes in supportive traits (Table 4.6). The primary criteria used to assess the economic value of the lines was milk yield through all lactations, length of time surviving and survival rates for three lactations. Differences in herd life and survival were small, particularly between the Milk and Index lines. The Young sire line averaged about two months lower in herd life probably because there was greater variation among their progeny. However the economic significance of the differences is nil because semen from the young sires was least expensive. It seems clear that as long as the best bulls are chosen by either of the criteria used and semen costs are considered, differences in profitability from the three methods will be small. Thus considering semen cost per unit of milk, selection for superiority in milk production is recommended. Findings with lines of Jerseys at Randleigh Farm of NC State University and a research station, Lewisburg, Tennessee are similar to that for Holsteins.

For progeny tested sires, the reliability for estimates of sires should be 65% or greater for PTA milk. High repeatability results from a large number of progeny in several herds; a wide range of herd conditions. Evidence shows such sires produce good "workable cows." PTA milk for these sires is generally more in the medium range, +300 to +600 kg. Choosing sires with PTA milk > +1200 kg is discouraged to produce replacement heifers. If the repeatability of the "high sire" is > 50%, the price of semen is high for general breeding. Sires with < 50% repeatability are not recommended as they generally have small

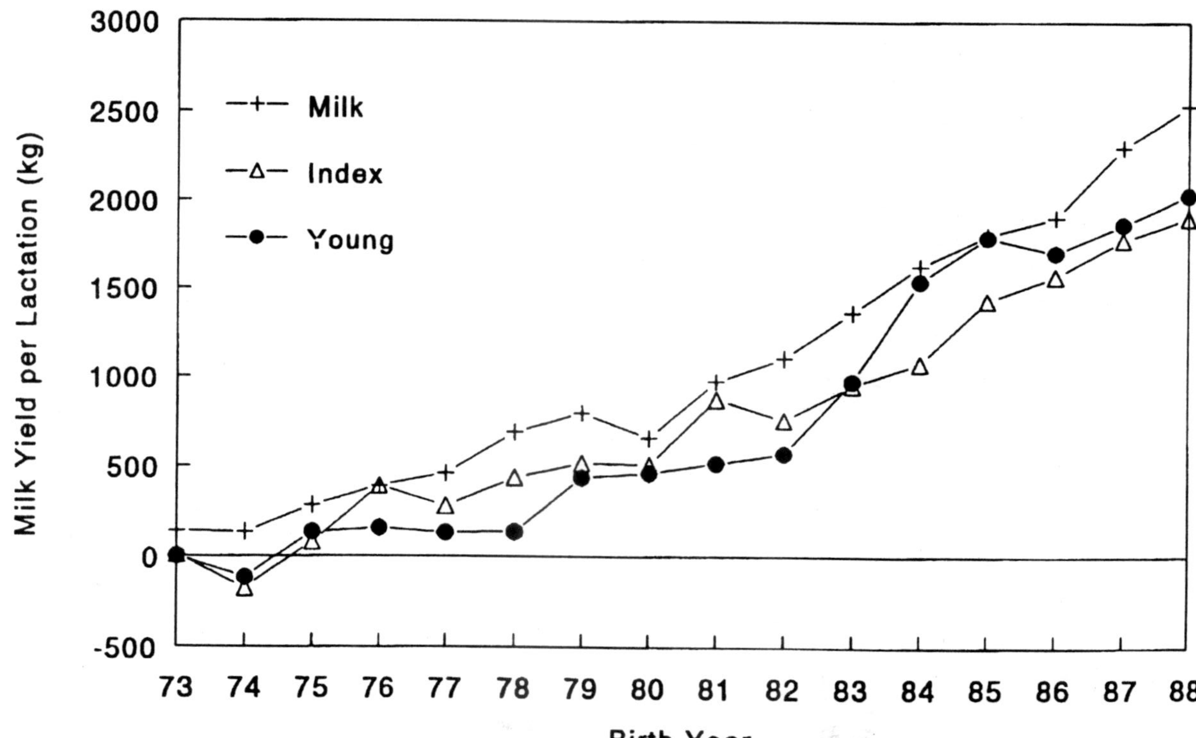

Figure 4.1. Averages for breeding values in NCDA herds computed from all lactations by line and birth year. Source: Adapted from McDaniel and Bell, 1992.

Table 4.6. Performance of three lines of Holsteins in herds of North Carolina during first lactation following four generations of sire selection.

	Lines		
Trait	Milk[a]	Index[b]	Pedigree[c]
Age 1st calving (mo)	27.8	27.8	27.6
Days in milk	277	276	275
Milk yield (kg)	8,053	7,821	7,716
3.7% fat corrected milk (kg)	7,815	7,682	7,619
Fat yield (kg)	186	184	186
% fat	3.71	3.76	3.79
Body weight (kg)	511	517	511
Heart girth (cm)	186	187	186
Days open	137	129	130
Days dry lactation 1-2	60	60	59
Herdlife (mo)	58.0	57.4	56.0
Cows surviving to:			
Second lactation	74.3	72.8	70.6
Third lactation	52.2	51.7	47.2

[a] Progeny tested sires chosen from all active AI bulls in the U.S. with repeatabilities of 60% or more and on level of PD for 3.7% FCM.

[b] Progeny tested sires indexed from an index combining milk, udders, feet and legs with 2 times as much emphasis for milk as for type traits.

[c] Young sires entering AI service based on pedigree merit for 3.7% FCM computed as 0.5 (sires predicted difference for milk yield + dam's cow index).

Source: McDaniel and Bell, 1992.

numbers of progeny in few herds (low range of environmental conditions). In Mexico, Colombia and Puerto Rico, progeny of sires medium to high in PD milk but < 50% in repeatability tended to be lower in milk yield and significantly lower than well tested sires in stayability or length of herd life (Abubakar et al., 1987a, Romero et al., 1989).

Small population size (< 200,000 cows) hinders the usefulness and cost effectiveness of progeny testing as a primary investment to enhance genetic gain. Should countries of WCZ choose to focus on within-country sire selection, the greatest potential economy is likely from exploiting the genetic pathway through dams of sires which will require systematic herd recording (Wiggans and McDowell, 1993). The potential for genetic gain in milk is no more than 100 kg per year. However, genetic gain is only achievable when combined with a good selection program (McDaniel and Dentine, 1985). The upper limit of gain through pedigree selection is 71% while progeny testing can be accurate to 99% in selection for milk yield. The cost of testing each bull retained after progeny testing is approximately $150,000, thus selection programs to ensure accuracy can be quite expensive.

It is recommended that selection of sires to introduce to a country become the responsibility of a committee of at least four members, with rotating membership and 50% of the members should be dairy farm operators. It has been shown time and again that herd owner participation in AI programs is highest when their peers are included in sire selections for importation of semen. Table 4.7 has the predicted transmitting abilities (PTA) for U.S.$ (dollars) values of protein, milk, fat and the reliability of the PTA's (Rel), type score, calving ease and semen cost for 10 sires offered by two U.S. AI studs. All sires are predicted to increase lactation yields of protein, milk and fat, but four are expected to lower percent protein and five to decrease fat percent. In type score (PTAT) all sires show an increase and therefore rather high TPI's (type production indexes). When the 10 sires are bred to heifers the estimated percent of difficult births (% DBH) ranges from 7 to 15. For cost of semen, the H (high) will be expensive for general breeding particularly when two or more services are required on average for conception. The M bulls (medium) could be used across the herd but with prices 50 to 100% greater than the L bulls (low price). Is the gain from use of M bulls worth the additional cost? The four L bulls are high in Rel with two positive in all PTA's. The other two show declines in %F and one in %P. As a member of the sire selection committee, which sires would be recommended for importation?

Table 4.7. Sampling of progeny tested Holstein sires on offer by two US AI studs.

Sire	Protein				Milk and fat					Type Score			Calving base		Semen
Code	Rel	MFP$	kg P	% P	Rel	MF$	kg M	kg F	% F	Rel	PTAT	TPI	Rel	%DBH	Cost
1	91	320	33	-.10	91	356	1380	38	-.12	83	.02	935	99	11	H
2	72	314	32	-.04	72	331	1173	49	.07	64	1.23	1116	68	15	H
3	84	293	28	-.15	84	345	1392	30	-.19	79	.99	952	70	13	M
4	64	262	25	-.13	72	305	1191	32	-.11	73	.68	855	36	10	M
5	85	260	32	.09	85	235	756	44	.18	82	.21	1060	98	10	H
6	73	254	29	.00	73	259	938	35	.02	73	.76	958	94	9	L
7	89	230	27	.03	90	226	808	32	.04	88	1.55	1096	97	8	L
8	99	228	28	.03	99	224	821	30	.00	99	2.21	1159	99	8	M
9	97	227	29	.00	97	231	947	19	-.16	93	1.59	1055	98	9	L
10	87	222	21	-.11	88	259	1029	25	-.13	80	.74	850	96	7	L

Rel - Reliability indicates accuracy of genetic evaluations.

MFP$ - weights PTA for milk, fat and protein to reflect gross income per lactation of progeny in US$.

kg P, %P, kg M, kg F and % F - PTA's for kg protein, % protein, kg milk, kg fat, and % fat indicate how much more performance to expect from an average daughter of a bull with a PTA of zero for the same trait.

MF$ - PTA in milk-fat $ weights PTA milk and fat to reflect gross income per lactation that future mature daughters will earn in excess of herdmates sired by bulls MF$ zero.

PTAT - PTA type is expected difference in final score between daughters of the bull and breed average.

TPI - Type production index by placing emphasis of 3 for PTA protein, 1 for PTA fat, 1 for PTA type and 1 for udder composite traits.

% DBH, estimate of percent of difficult births in heifers when bred to given bull; use when selecting bulls to breed heifers.

Semen cost: H, high; M, medium, L, low.

Country Policy for Selective Breeding

The thrust in the discussion thus far has been developing a rationale for producing and maintaining cattle with sufficient genetic quality to support commercial dairy operations in countries of the WCZ. At the same time there were attempts to dispel some of the perceptions on breeding plans which are not supportable at this time, e.g., selective breeding for high adaptation to warm climates can be genetically correlated with high milk yield. These are all features for consideration but when governments need foreign exchange from other sectors of the national economy to pay for imports of semen or embryos and governments are also promoting self sufficiency in total agriculture production, the outflow of currency for dairy operators in purchasing semen or embryos on a continuous basis will be challenged.

Although Holman et al. (1990) showed that semen from AI progeny tested sires from abroad could be used profitably in Columbia, Mexico and Venezuela, they suggested importing pedigree selected young sires to produce semen locally for distribution at lower cost. Costs of importations of semen to breed the national dairy herd is an issue in Zimbabwe. Mpofu et al. (1993) developed computer models and evaluated nine proposed plans or systems for supporting national plans for genetic gain in bulls and cows. The objective was to determine the economic merit of nine breeding strategies. Their nine proposed systems, and estimates of achievable gains in milk yield and annual costs in local currency over 25 years are given in Box 4.1.

A locally executed progeny testing scheme (PT1, Box 4.1) was lowest in estimated cost. There is some tenuity in the sum of 1106 (10^3) as being total cost for PT1 since there were no capital costs for infrastructure included and only part of the operating costs on record processing and for qualifications of needed personnel to execute effectively. Use of progeny testing as in PT1 but with imports for sires of sons (PT3) was second lowest in cost. Similar to the PT1 system, there may be some unaccounted costs, nevertheless PT3 could be a worthy approach as the achievable response of 1.51 SD is better than for PT1. The system CS 130 (importing semen to breed 30% of the cow population) is above average in cost and lowest in expected results, low SD. The ELITE System, use of untested young bulls produced from foreign elite bulls and local elite cows, is certainly worthy for consideration as a national program. This system is projected as above average in expected gain (1.70 SD) and has the added potential of motivation on the part of local producers for breed improvement when their cows are selected as bull dams.

Which approach becomes adapted by Zimbabwe continues unsettled; nonetheless well worth the effort in gaining support locally for a national dairy cattle improvement program in genetics.

Choice of Breed

In the temperate areas the major breeds used most widely for dairying are: Brown Swiss, Guernsey, Holstein or Friesian and Jersey. These are all recognized worldwide. There are numerous other breeds found largely within a region or country, such as the Red Dane of Denmark, Norwegian Red in Norway, Swedish Red and White in Sweden, Black Pied in Russia, and Simmental of Eastern Europe. There have been samplings of all the main breeds as well as from the smaller numbered in the WCZ. By far the greatest in number both in the temperate areas and the WCZ is the Holstein followed by Jersey but in much fewer numbers.

There are herds of good quality cattle in all breeds. However where numbers are small, there are certain cautions recommended. For instance, the selection process in a low population breed may be dominated or controlled by a few breeders with their own criteria. Phenotypic characteristics such as hair coat color and body conformation will often receive priority over maximizing milk yields. Another factor is that inbreeding effects can rise to the point that progress for increasing milk yield can be small or nil. With large differences in economic or environmental conditions, breeds may differ in ranks for total usefulness (McDaniel and Dentine, 1985).

System Code	Strategy	Achievable Response 25yr[a]	Annual cost ($2x10^3)[b]
PT1	Closed progeny - testing scheme in Zimbabwe.	1.06	1,106
PT2	Progeny - testing with 30% foreign bulls as sires of cows.	1.65	5,038
PT3	Progeny - testing with foreign bulls as sires of all bulls.	1.51	1,446
NMOET[c]	Closed MOET nucleus breeding scheme based initially on elite imports.	1.75	9,159
CSI	Continual semen importation for 100% of cow population.	2.23	4,195
CSI30	Continual importation for 30% of cows.	.89	6,831
CSI50	Continual importation for 50% of cows.	1.39	13,423
ELITE	Young bull team out of foreign elite bulls and local elite cows; bulls used untested.	1.70	1,767
CEI	Young bull team out of imported elite embryos; bull used untested.	2.03	3,771

[a] Units are standard deviations (SD) taken as half the theoretical response; SD = 540kg.
[b] Mean annual costs over 25yr.
[c] NMOET - closed multiple ovulation and embryo transfer nucleus breeding scheme, based initially on elite imports.
Source: Adapted from Mpofu, 1993.

None of the dairy breeds are superior in all characteristics which can create economic variability in a dairy enterprise. Hence in choosing a breed or developing a herd breeding program it is wise to consider a number of traits taking market conditions into special consideration. As an example, Box 4.2 gives the relative comparison of the Jersey breed to herdmates of other breeds around the world. The Jersey shows comparative advantage in a number of characteristics, such as in milk composition, age to reach puberty, use of forages, resistance to

Box 4.2. Breed characteristics that can create economic variability for dairy enterprises with Jerseys as an example.

Characteristic	Indication relative to other breeds
Growth	
Birth weight	Low but good calf vigor
Heifers	Among earliest in puberty and breeding
Skeletal dimensions	Good rate of frame development
Postpartum	Weight loss low, rapid return to energy balance and estrous cycle
Rate of maturity	Among fastest breeds
Milk yield	Lower than Holsteins but above most breeds
Mobilizing body reserves	Capability among highest
Persistency of milk yield	Above average
Milking rate	Average or slightly below
Milk constituents	
Colostrum	Excellent quality; protein to fat ratio could be low for Jersey calves
Fat characteristics	Excellent in most markets
Protein	High to highest in percent and quality
Minerals	Breeds appear similar
Cheese production	Best for quality cheeses and yields
Health	
Mastitis	Low in clinical mastitis
Milk fever	Among the high risk breeds
Udder edema	Breed is average
Displaced abomasum	Breed effects negligible
Laminitis	Lower risk than most breeds
Stayability	Among best post first calving but not in % born completing 2 lactations
Calf mortality	Undesirably high
Age at puberty	Has comparative advantage
Age at first calving	On average has comparative advantage
Gestation length	Among breeds with lowest days
Calving interval	Among best breeds with advantages in certain environments

Characteristic	Indication relative to other breeds
Calving ease and dystocia	Among best of all breeds
Lactation length	Average among dairy breeds
Breeding efficiency	Above average; losses from sterility influenced by environment
Udder characteristics	
Teat size	Smaller than most other breeds
Teat placement	Among best breeds
Udder volume	May have slightly lower secretory rate per unit of tissue than Holstein due to higher fat and protein
Udder strength	Well above average, overall high quality udders
Blood constituents	Breed differences usually quite small; may have merit in low leukocyte count
Nutrition	Shows advantages; excellent user of forages
Body composition	Deposits excess abdominal fat, excellent capability to mobilize fat reserves
Temperature stress	Highest in tolerance for dairy breeds
Biochemical polymorphisms	May have unique hemoglobins but among highest in autosomal recessive gene frequency
Semen characteristics	May have comparative advantages in quality post thaw motility
Economics	Comparative advantages especially in cheese markets
Beef production	
Pure Jersey	Low performance in intensive systems; on grazing alone may equal or exceed several breeds in leanness and meat quality
Jersey sired crosses	No comparative advantage
Jersey as dam	Highly regarded as dam for terminal crossing; medium in beef yield
Jersey cross dam	Especially good mothers

Box 4.2. (continued)

Characteristic	Indication relative to other breeds
Crossing for milk production	
Jersey as sire	Below Holstein crosses in milk; low environments net indexes high
Jersey as dam	Produces excellent crosses with good efficiency
Coat and hair	Among best breeds for tolerance to environmental conditions
Physiological characteristics	Has some unique qualities
Variability in condition score	Responds above average among breeds
Temperament	Excellent, one of best breeds
Draft power	Crosses preferred in several countries for small farm agricultural power
Selection	
Heritability	Estimates lie within range of all dairy breeds
Repeatability of lactation yield	Slightly above average for dairy breeds
Results from selection	Responds to selection for milk and composition as readily as other breeds
Effectiveness of selection programs	Jerseys have excellent programs underway in at least 3 dairy countries (Denmark, New Zealand and U.S.)

Adapted from McDowell, 1988.

mastitis, and calving ease; is average among breeds in most traits; but has limitations in a few traits just as any other breed. Evaluation of the indications in Box 4.1 leads to the conclusion that the Jersey breed has the potential for greater utility in areas of the WCZ than is occurring at present, not only as a pure breed but also has high utility for crossbreeding programs.

Similar profiles could be assembled for other breeds. For instance, Holsteins would show comparative advantages in yields of total milk, fat, protein and total milk solids, mature body size, availability of progeny tested sires and potential in beef production; be average in breeding efficiency, duration of herd life, udder characteristics and calf mortality; but be marginal in milk composition for numerous markets, e.g., yield in cheese making, problems at first parturition unless well grown out and is not the most efficient user of forages.

A scenario to consider in a breeding program in WCZ would be combining Holsteins and Jerseys as crosses for best achievable economic and biological efficiencies. Both breeds have shown good combining ability with each other and with additional breeds (McDowell, 1982); response to selection for milk and milk composition has transpired more readily than for most breeds; and both breeds have effective selection programs in action, e.g., Jerseys in Denmark, New Zealand, Australia, South Africa and the U.S. as well as at least 8 countries for Holsteins. These two breeds have also contributed effectively to programs in beef production and draft power.

Pursuit of the technical literature leads to the prediction that overall the most effective cow for many environments in the WCZ could be from a crisscrossing breeding program between Holsteins and Jerseys oscillating around ⅜ to ⅝ genes per breed. Level of heterosis in the crisscross cow would be lower than for the F_1 cross but heterosis should be sufficient to aid in early age at puberty, rate of rebreeding postpartum and some in resistance to health problems. Expected milk composition would have added acceptance in many markets.

The central points on choice of breed are to direct awareness to the complexity of dairying and that selection should involve a number of traits. There are comparative advantages among breeds which could be used to develop gene pools to provide gains for dairying in the WCZ.

CHAPTER V

BREEDING EFFICIENCY

Background

As long as various species of animals maintain visibility of presence, rate of reproduction is of no concern. But as humans have become dependent on rapid turnovers in domestic mammals and poultry, rate of reproduction has become a vital supportive trait in maintaining desired economic returns. For instance, a cow completes a lactation; the profitability from the sale of her milk and that from the entire herd depends on the proportion of the milk credited as annual income (MC/yr) calculated as days per year divided by calving interval (CI) and multiplied by milk yield (MC/yr = 365d/dCI x milk yield). Our goal is that cows will lactate for 10 months, have a dry period of 60 days prepartum for rest then recalve in 365 days. This means all the lactation milk is credited as annual income. CI can markedly affect the annual rate of return as shown in Table 5.1.

For Holsteins in Zimbabwe the mean CI across herds is 407 days, the cows lactate 291 days and lactation milk averages 5,860 kg. The CI ratio reduces the annual credit by 731 kg or to 5,136 kg, which is a significant loss. The 365 days per CI ratio ranged from 65 to 98% among herds indicating that herd averages per year range from 5,743 kg to a low of 3,801 kg due to rebreeding inefficiency (Makuza, 1993). Low milk credit is the "short term" effect but as CI rises "long-term risks" increase. The average days dry in Zimbabwe is 116 days which is 56 days above optimum in non-productive time and adds to feed cost per cow. When cows have dry periods exceeding 100 days in the WCZ, there is a high probability of problems at parturition, such as retained placenta, frequent metabolic disorders, such as ketosis, and fat cow syndrome, all of which will affect subsequent lactation milk yield and rate of rebreeding.

Table 5.1. Influence of rate of rebreeding on creditable annual milk sales per cow.

Pregnancy postcalving (d)	Calving interval (d)	Credit per year (%)	Lactation Milk (kg)		
			4000	5000	6000
86	365	100	4000	5000	6000
115	395	92	3680	4600	5520
145	425	85	3400	4250	5100
175	455	80	3200	4000	4800
205	485	75	3000	3750	4500
235	515	71	2840	3550	4260
265	545	67	2680	3350	4020
295	575	63	2520	3150	3780

In Canada, the U.S. and Western Europe above 40 °N latitude, CI is usually between 380 and 398 days. The values for CI in Table 5.2 indicate Holstein herds in the lower latitudes (WCZ) average 9 to 15% longer in CI. The values are from over 1000 herds on milk recording which means they likely average shorter in CI than most herds. There are month of calving effects on CI ranging from about 4% in Colombia and Zimbabwe to 10% or greater in the other locations. The longest intervals in Mexico, Colombia and Taiwan follow calvings from September to November which is attributed to the decline in feed quality in mid-winter. In Puerto Rico and North Carolina the longest intervals are for cows calving April to June. High temperatures July to September suppress return to estrus and for conception in these cows. Older cows appear to have somewhat longer CI as indicated by differences between intervals during first lactation and later lactations.

When level of lactation milk is high, most cows that are slow in rebreeding, such as in Mexico and North Carolina, can continue milking profitably beyond the usual 10-month period. Such would be feasible for only a few cows in herds of Colombia and Puerto Rico (Table 5.2). When milk yield is low, a downward spiral is almost inevitable. Income is low for investment in remedial measures, such as more feed, to improve breeding efficiency, hence the herd ceases to operate or continues but at high economic losses. Even in the higher producing herds, owners claim they can ill afford to support the labor required for nearly continuous observation for estrus. The remaining discussion will feature possible manipulations of herd management and feeding practices to maintain good rebreeding efficiency.

Factors Influencing Breeding Efficiency

Numerous factors can be associated with the rate of conception, some of which result from genetic limitations but by far greater are environmental effects (Box 5.1).

1. <u>Genetic Limitations</u>

It is generally accepted that the heritability of traits used to assess breeding efficiency is too low for incorporation in a selection index but there may be some change with intensity of selection for milk yield. Data gathered prior to 1960 gave heritability estimates of breeding efficiency of -0.06 but records since 1970 average 0.06 to 0.12 in heritability. This rise indicates sire variance for breeding efficiency may be increasing. There are also some anatomical, physiological and functional defects affecting rebreeding rate. Possible cytoplasmic effects on reproduction is gaining in attention as well. In Holstein herds of North Carolina, cytoplasmic effects range from slightly larger than additive genetic effects to twice as large. Hopefully, time will yield more conclusive evidence on cow line effects on reproduction.

There is evidence for higher frequencies of the incidence of stillbirths (calf born dead) and dystocia (difficult parturition) in European breeds than Zebu types or *Bos taurus* types with adaptive history in the WCZ. Data on Holsteins from herds in the WCZ are as yet difficult to evaluate due to different definitions of stillbirth, breed of sire effects, and the influence of poor feeding. Thus far environmental effects have proven most significant in stillbirths.

Table 5.2. Calving intervals by month of calving for Holsteins in commercial dairying of various countries.

Calving month	All parities				Puerto Rico		North Carolina	
	Mexico	Columbia	Taiwan	Zimbabwe	1st-2nd	Later	1st-2nd	Later
Jan	409	427	411	412	421	427	410	419
Feb	403	418	402	408	427	431	427	432
Mar	401	422	423	408	429	434	443	438
Apr	402	414	421	406	433	441	441	453
May	404	420	420	407	442	444	442	444
June	413	416	419	401	445	441	445	441
July	422	409	405	403	408	425	408	431
Aug	424	425	406	402	410	413	410	420
Sept	434	425	441	407	401	409	401	410
Oct	433	429	449	412	411	411	411	408
Nov	430	431	413	413	409	412	410	416
Dec	418	422	401	409	417	421	402	409
Av.	416	422	418	407	421	426	421	427
<u>Milk credit/yr</u>								
(%)	87.7	86.4	87.3	89.7	86.6	85.6	86.6	85.4
<u>Av. milk yield</u>								
(kg)	7156	4771	6207	5860	5464		7528	

┌───┐

Box 5.1. Factors affecting breeding efficiency in WCZ.

Factor	Some Effects
Genetic limitations	Some functional defects
Thermal stress (hours $> 27°C$)	Day-night differential important in estrus and conception
Nutrition	
Low protein	Delays puberty
Low energy	Delays puberty and time of estrus after parturition
Mineral imbalance	Especially phosphorus, delays breeding
Over fattening	Slows conception and results in excessive weight loss in subsequent lactation, delayed breeding
Loss of body weight	Cow must be in steady state or gaining in weight
Health	
Disease	High body temperature prevents conception
Internal parasites	May cause sterility
Foot problems	Delays movement and expression of estrus
Level of milk yield	In WCZ each rise of 1,000 kg above 3,000 kg will increase days open approximately 10 unless feed level rises
Parity number	Calving interval between first and second calvings longest; also old cows (>7 years) have longer calving intervals
Dystocia	Low when cows are adequate in size but can be quite high in low weight heifers
Herd size	Negative correlation between herd size and breeding efficiency
Degree of confinement	Close confinement can markedly influence estrous detection
Lot surface	Wet concrete floors lower breeding efficiency
Social order	Shy cows are slow breeders; must be watched
Space per animal	Desirable to have cows on dirt surface with 30 m^2 per cow for at least 2 h per day
Estrous detection	More difficult in WCZ
Drug therapies	Not generally recommended for treatment of anestrus due to suppressed thyroid output

└───┘

Breed differences, age of cow, size of cow, and breed of sire appear more important in problems of dystocia than differences among cows within breed. Guidelines have become reasonably clear on means to minimize the incidence of dystocia, such as sire selection to breed heifers as indicated from the index values in Table 4.6. However, the incidence of dystocia among cattle breeds from the temperate areas (large breeds) appears higher than breeds native to the WCZ (small breeds). At first parturition, Holsteins should have a body weight equivalent to 85% of expected mature weight (450 kg humid tropics vs. 550 kg for temperate areas) to minimize dystocia. The frequency of dystocia rises dramatically for heifers weighing only 75% of expected mature weight (McDowell, 1987). Correa et al. (1993) showed through pathway analysis that associative effects may rise. For example, dystocia increases the occurrence of retained placenta as well as the odds for metritis. Stillbirth increases odds for mastitis and retained placenta. The odds for dystocia and mastitis are lower following birth of females than males with twin births having highest risks. Dystocia at parturition, retained placenta or ketosis increases the odds for metritis. The conclusion is that stress from one observed cause leads to an increase in the incidence of other disorders.

In well fed herds about 33% of calvings in Holsteins need assistance; about 22% require an easy pull, 8% a hard pull and 3 to 4% require veterinary assistance. These figures are much higher than would be usual in Jerseys, for example. Calf mortality is associated with the level of difficulty at delivery. The mortality rate often rises when assistance is given because farm personnel are "too quick" to pull and do not work with the cow as contractions occur. Decisions on assistance should be made by responsible, experienced personnel. Unless the calf is breached, assistance should not be given until the calf's tongue starts to swell or turn color. Then use only sanitized equipment when pulling a calf. If the placenta is not released within 6 hours of calving, the cow should receive oxytocin or prostaglandin. Manual assistance for removal of the placenta is not recommended (Ferreiva, 1991).

The onset of puberty is genetically determined and conditioned by various exteroceptive influences. Females of most improved dairy breeds may show signs of heat as early as 7 months and reach puberty both chronologically and physiologically earlier than most breeds of Zebu; 325 days for Jerseys and 400 days for Holsteins vs. 600 to 780 days for Zebu. Bos taurus types native to the WCZ, such as Criollo cattle of Latin America, are a bit later in reaching puberty (480 days) than breeds in temperate areas (Lemka et al., 1973), but are much younger than Zebus in the WCZ. Although age of first calving is not a measure of later reproductive efficiency, it is a significant factor from an economic standpoint. Correlations between age at first calving and subsequent calving intervals are low (-0.11 ± 0.07) in both European breeds and Zebu which indicates independence between the two traits. This means inherent differences among breeds for age of puberty has little relation to reproductive efficiency after first parturition.

Another factor which can be important in the use of improved dairy breeds is some association of genotype and a condition described as "wooly coat" (long hair with ends curled over). We know the condition can arise from poor nutrition, internal parasite infestation and diseases, but there may be genetic linkages as in Latin America, some sires had 15% wooly coated progeny, 60% intermediates and 25% short haired while others produced progeny with only intermediate and short hairs (McDowell, 1972). The condition wooly coat can create problems in functional efficiency because the hair coat affects responses to heat stress. Culling of wooly coated animals is recommended as milk yields will be well below herd average.

From present evidence, anatomical and physiological defects which depress reproduction are not likely to occur with any greater frequency among improved dairy breeds than other genotypes in the WCZ.

2. Thermal Stress

Irrespective of environment, cattle have AM and PM periods of high activity (Figure 2.3). As maximum daily temperature rises, occurrence of peaks in eating and activity become more distinct; therefore, behavior patterns become important in developing management practices to maintain good breeding efficiency. For example, the time of estrous behavior is distributed from throughout the day in New York but is concentrated early AM and late PM in Venezuela (Table 5.3).

Table 5.3. Distribution of mounting by time of day in a hot climate (Venezuela) vs. cold (New York State).

	Time	Percent of Total	
		Venezuela	New York
AM	1:00 - 7:00	54	43
	7:00 - 12:00	7	22
PM	1:00 - 7:00	3	10
	7:00 - 12:00	36	25

Source: Adapted from McDowell, 1972.

In Holstein herds of North Carolina, month of calving is also related to frequency of incomplete lactations with highest occurrence in first lactation (Table 5.4). High losses in April and May reflect culling for slow rebreeding during the hot summer while the losses for September calvings appear related to health problems, largely laminitis in the rear feet. This resulted from acidosis following reduced intake of forages (hay or grazing) and use of higher energy density feeds to mediate the effects of heat stress during the early months of lactation. Overall, the percentage of incomplete lactations is less during second lactation but June and July calvers also appear susceptible to metabolic disturbances as they attempt to combat thermal stress.

The use of shades or shelters for lactating cows can yield benefits in the reduction of thermal stress and maintaining rebreeding efficiency (See Chapter VIII, Housing). Cows continuously confined to sheltered areas show no benefit in feed intake or frequency of estrus over non-sheltered cows. Keeping cows inside a barn or shelter from 10:00 AM to 4:00 PM

Table 5.4. Influence of month of calving on the percentage of incomplete lactations in Holstein herds of North Carolina.

	% Incomplete	
Month	1st lact.	2nd lact.
Jan.	17.3	9.0
Feb.	14.3	8.7
Mar.	17.5	7.1
Apr.	25.2	15.5
May	23.8	16.6
June	18.2	18.2
July	21.0	20.9
Aug.	22.8	10.8
Sept.	25.7	16.9
Oct.	19.0	17.9
Nov.	14.3	9.5
Dec.	16.0	11.7

Source: Adapted from McDaniel et al., 1986.

followed by lock out into open lots with a dirt surface the rest of the time is more satisfactory than free choice in the use of shelters. Moving cows from inside to outside stimulates activity which enables better detection of estrus (McDowell and Leining, 1978). In six large commercial dairy herds of Puerto Rico, calving interval decreased from 432 to 400 days with the adoption of mid-day shelter followed by dirt lot or pasture from 4:30 PM to 2:00 AM and 5:00 AM to 10:30 AM. The key is to observe at these times of movement. Heifers also need daily exercise to stimulate estrus and produce good size follicles.

3. <u>Nutrition</u>

It is generally recognized there is a positive relationship between feeding level and reproductive efficiency, but the effects on reproduction remain only partially understood due to considerable variability in results from designed experiments. For instance, the number of follicles may not be affected by nutrition but nutrition level does affect follicle size with estradiol concentration twice that of low fed heifers. Cows on low carotene levels appear to have higher incidence of retained placenta. It has also been shown there is a likely relation between deficiency in selenium and vitamin E and retained placenta. Although there are claims for better breeding efficiency following injection of vitamin A, the majority of results from experiments in cattle have shown no significant response. There is some support for vitamin D supplementation intensifying the expression of estrus but the results are inconsistent.

Some have stated that conception will improve in heifers when supplemented with phosphorus while others report no benefits. Apparently when there is adequate calcium for a positive balance early in lactation additional phosphorus may be required. Obviously, the effects of minerals and vitamins on reproduction remain unclear. The effectiveness of mineral and vitamin supplementation may therefore depend on quality of feeds, especially forages, and stage of gestation or lactation.

Heifers of improved dairy breeds should have about one centimeter depth of fat deposit showing over the body at time of parturition. This represents a condition score of 3.5 (Figures 5.1, 5.2). Higher levels of fat (score 4 or 5) are associated with postpartum metabolic disorders. A similar recommendation is made for cows. It is difficult to perceive in the WCZ, where a primary target is for more feed, that over conditioning occurs more frequently than recognized. This is because long calving intervals and long dry periods often cause fattening.

This brief outline on problems of nutrition shows strong need for investigations on management of feed resources to achieve acceptable breeding efficiency.

4. Level of Milk Yield and Body Weight Changes

Ovarian activity is low if females are losing weight and condition. Body weight status of cows markedly influences rate of conception during the first 100 days following parturition. Conception rate is higher and services per conception lower for cows gaining in body weight during the month of rebreeding (Table 5.5). As pointed out in Chapters II and III, the suppression of appetite by heat stress makes it difficult to get body weight adjusted for early rebreeding. Weight loss and delay

Table 5.5. Influence of early postpartum change in body weight on rate of conception in Holsteins in the US.

	Weight status in month of breeding	
	Gaining	Loosing
Total services	1368	544
Pregnancies	911	234
Conception rate (%)	67	44
Services/conception	1.50	2.32

Source: McDowell and Leining, 1978.

in the onset of estrus may be associated with changes in net energy balance during lactation. Figure 5.3 illustrates the net energy balance (MCAL/d) of a lactating Holstein with high milk yield. She was in a serious negative state of energy balance (curve below "0") for about 12 weeks even though feed was offered free choice. During weeks 1 to 8, 55% of her milk came from the utilization of 1.82 kg of her body stores per day. After week 32, energy balance was positive and the cow gained weight. Her energy balance was in equilibrium by week 13 and it was then that ovarian activities returned to normal and 21-day estrous cycles commenced. A central point is that careful observation of lactating cows for signs of stability of weight can yield dividends.

Another way to view energy balance and rebreeding is a possible negative relation to level of milk yield and days open as shown in Figure 5.4.   Days open of cows producing in the highest quartile (1,000 kg above herd average milk yield) was 37 days higher than the lowest quartile. McDowell et al. (1976) concluded that dairymen in Mexico and Puerto Rico gave emphasis to type score in the selection of sires because they felt that although the progeny might have lower milk yields than the highest milk bulls, breeding efficiency in their view would be better from shorter periods of body weight loss.

Cows may return to estrus sooner during first lactation than in later parities but they are usually less fertile. It is therefore recommended that time of rebreeding commence at 45 days

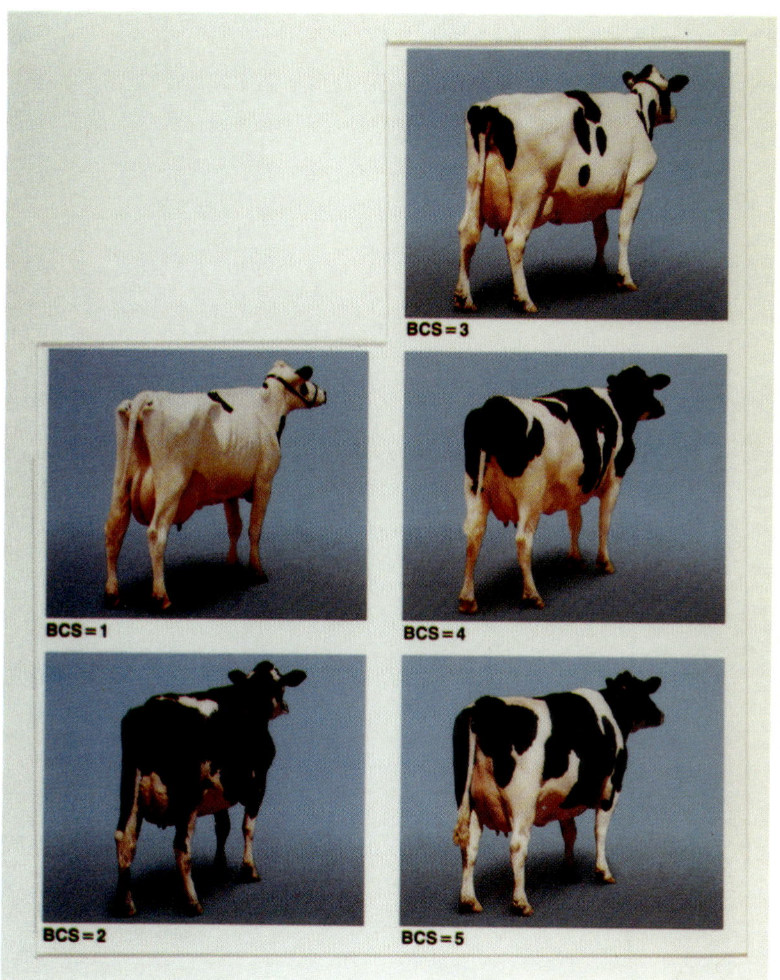

Figure 5.1. In dairy cows, body condition is an indicator of the amount of stored energy reserves which changes with different stages of lactation, therefore scoring can be an effective procedure for feeding, reproduction and health; BCS1, too thin, no reserve; BCS2, usually early lactation; BCS3, mid-lactation; BCS4, end of lactation and dry period; BCS5, overly fat, not desirable (Courtesy of Elanco Products Company, Division of Eli Lilly and Company).

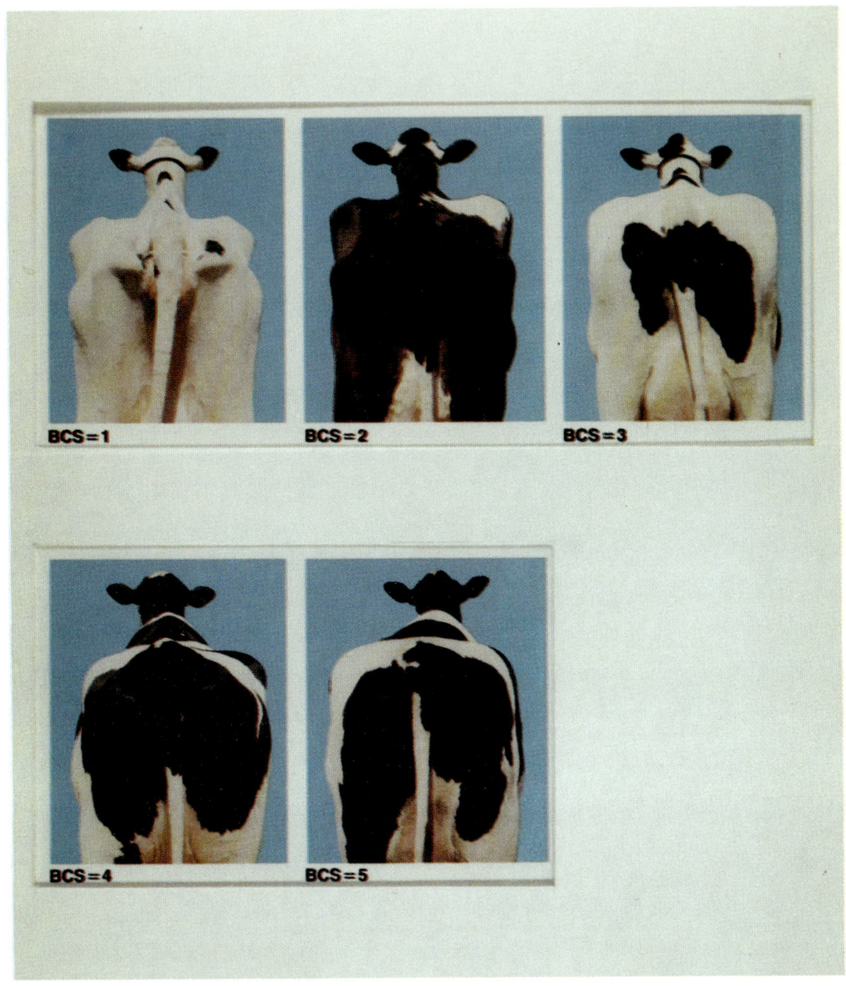

Figure 5.2. Body condition scoring from rear: BCS1, deep cavity around tail head, no fatty tissue in pelvic or loin area; BCS2, shallow cavity around tail head; BCS3, no cavity around tail head, fatty tissue easily felt over whole area; BCS4, folds of fatty tissue around tail head with patches of fat covering pin bones; BCS5, tail head buried in thick layer of fatty tissue, pelvic bones cannot be felt under firm pressure (Courtesy of Elanco Products Company, Division of Eli Lilly Company)

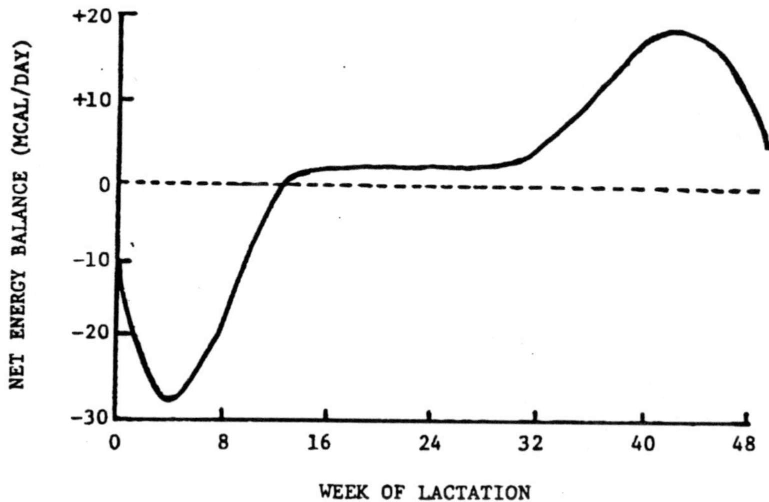

Figure 5.3. Net energy balance of high yielding dairy cow during 48-week lactation period (Source: Flatt, 1965).

MILK YIELD & DAYS OPEN

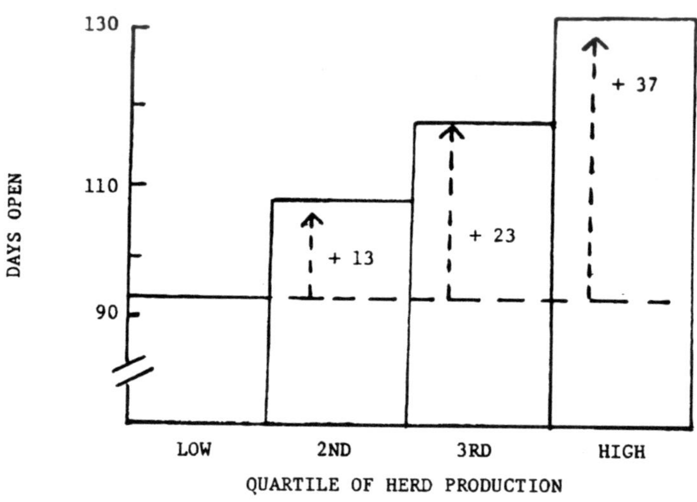

Figure 5.4. Milk yield and days open by quartile of herd production of Holsteins (Source: Spalding et al., 1975).

postpartum. In fact, because lactation periods tend to be less than 10 months in the WCZ, it is practical for all breedings to commence at the first estrus following the 45th day of lactation. Cows treated for metritis or other conditions should be cleared of the condition before being bred.

5. Detection of Estrus

Although nutrition and lactation contribute to rebreeding rate, failure to detect estrus can extend the problems. Satisfactory heat detection requires time and skills (Box 5.2). The best single indicator of estrus is when a cow stands as mounted by another cow. Therefore, having willing mounting animals in the herd is important for success. For herds in the WCZ, the frequency of mounting is generally less due to lower stimulus for physical activity under heat stress. Some researchers have reported shorter duration of estrus under hot conditions but this evidently applies more to "visible duration" than actual hours. Observing for reddening and swelling of the vulva accompanied by a clear mucus discharge may serve to help identify estrus when the frequency of mounting is low.

Stage of lactation, freedom of movement and condition of the surface can all be important in detection and intensity of estrus (Box 5.3).

In temperate areas, over 80% of the estrus cycles are less than 24 days but in the WCZ, 30% or more of the cycles will range from 25 to 44 days. This means that cows or heifers projected for heat need longer surveillance.

As an aid in estrous detection, a written protocol is recommended in order that employees will have tangible material to follow. Senger (1990) suggests the protocol include:

- who will check for estrus
- how long will be needed to observe
- when to observe
- what signs or behavior patterns to look for
- which cow appeared in estrus and what was seen
- who will be notified and how (hand written and/or entry into a computer)
- how are records to be kept on estrous detection observations.

Box 5.2. Heat Detection.

Accurate heat detection takes time and skills. In
the WCZ up to 80% of detectable heats can be
observed from 9:00 PM to 7:00 AM which means
estrus detection should include observing during
the late evening and early morning. A single
observational period is not enough. Painting or
chalking tailheads is a low cost method and
generally is more effective than the rump attached
mount detectors, such as the Kamar Heat-Mount
Detector. Use of enzyme - linked immunosorbent
assays (ELISA) for progesterone can be accurate in
detecting estrus and ovulation. There are kits
available commercially for on farm use or milk
samples can be processed in a central laboratory.
Cows determined as not cycling can benefit from
two injections of prostaglandin F_2a 11 to 14 days
apart. Breeding should be made based on
observed behavior or clear vaginal mucus 80 hours
following the second injection. Insemination will
result in some pregnancies but at a lower
conception rate. However, in the WCZ,
prostaglandin should be used only on cows stable
(BCS 2 +) or gaining in weight. Where hired labor
is employed, there should be written instructions
on heat detection.

Source: Adapted from Ferreria, 1991.

Box 5.3. The Cow: Where I Best Show Estrus?		
Location	My Problem	Mounting Index
Milking parlor	space and wet	0.1
Feed bunk while eating	too busy	0.2
Holding area	too crowded	0.3
Dry concrete alley	space limiting	1.0
Dry concrete alley and movement	some risk of slipping	1.1
Dry dirt lot	good surface, need space	1.6
Dry dirt lot and movement	good movement	1.8

Times I stand and my duration of estrus (h)

Dirt Surface		Concrete Surface	
No.	Time (h)	No.	Time (h)
5.0	13.8	3.9	9.3

Source: Adapted from Stevenson, 1993.

The language or terms used in the protocol must be understood by the workers, e.g., workers employed in Thailand from Laos, or Spanish speakers in U.S. herds.

Another reason for close attention to estrous detection is to identify aborting females. Cows aborting <150 days after service may go undetected unless it is the practice to observe closely for discharge.

Commercial daily operators tend to judge the effectiveness of labor by cows milked per worker. When the worker feels there is a full time job with milking and other "visible result chores," little time is devoted to observing for estrus. In much of Latin America, dairy operators are offering incentive awards in money for reporting estrus which has proven to the advantage of all.

6. Reproductive Disorders

Almost any infectious agent that has a detectable effect on cows may interfere to some degree with the reproduction cycle, and from infections of the reproductive tract to nutritional disorders, such as diarrhea or direct effect of pathogenic conditions, such as foot rot. High infestations of endoparasites will delay puberty in heifers and poor development of the ovaries, "infantile ovaries." Chronic diarrhea from endoparasites in pregnant heifers can also lead to high losses of both cows and calves at parturition (McDowell, 1972).

The question of vaccination against brucellosis in the WCZ remains a debatable issue. Herd owners are cautioned to treat abortions as if they were due to brucellosis infection and take precautionary actions. Non-specific infections of the reproductive tract are becoming increasingly serious with the advent of larger herds in most all environments.

Factors which prevent uterine involution, such as secondary infections caused by retained fetal membranes and metritis upset the sequence of events in the normal cycle. Herd owners of all dairy breeds should hasten to use preventive and therapeutic treatments as milk yield may be depressed by 240 kg for retained fetal membranes, about 180 kg for dystocia, and 100 kg for metritis (Silva et al., 1992). Antibiotic therapy may seem attractive for treatment but when infusions are made routinely in a herd without regard to the stage of the estrus cycle, there will be detrimental effects on reproduction even though such treatments will not alter intervals from calving to first ovulation or conception. Among reasons for caution in antibiotics are that treatment with uterine boluses may introduce infection when infusion occurs during the progesterone phase of the cycle days 7 to 16. Infusions of some antibiotics between days 3 and 9 of the cycle will terminate the corpus luteum (CL) prematurely, while late infusion, day 14 to 17, prolongs the life of the CL. When used indiscriminately, intrauterine antibiotics can increase the variability of estrous cycles in normal cows thereby complicating breeding. On the other hand, intrauterine infusions may prove beneficial 24 hours following insemination especially if the heat detection program is adequate to determine the initial onset of estrus but if insemination occurs more than 20 hours after estrus is observed benefits may be nil. Guidelines on the merits and limits of some antimicrobials used to treat uterine infection are in Box 5.4. Because of risks with antibiotics, many veterinarians recommend flushing with Lugols solution in lieu of antibiotics. A further caution in the use of antibiotics or sulfonamides is that for several days following treatment the milk should be discarded as a precaution to human health.

Leptospirosis, campylobacteriosis, trichomoniasis and brucellosis are the major diseases affecting reproduction. Brucellosis can be controlled through calfhood vaccination and leptospirosis by periodic vaccination. The use of AI is recommended to prevent campylobacteriosis and trichomoniasis. Prevention and control are important as all four diseases can cause abortion; brucellosis during the last tri-semester of pregnancy, leptospirosis anytime and the other two in the first or middle third of pregnancy. Cows suspected as infected with any of these diseases should be isolated (Hoard's Dairyman, 1990).

Swelling and redness of the vagina, referred to as vaginitis, is not attributable to a specific microorganism. The condition may occur at any time but most frequently accompanies a uterine infection like metritis following a retained placenta. A white grayish or yellow discharge indicates the infection is chronic. Recommended therapy is with antibiotics. AI should be with a guarded pipette to avoid moving the infection further into the reproductive tract. Prevention and control of reproductive disorders, including diseases, are quite important aspects of herd reproductive efficiency.

7. Cystic Ovaries

The condition of cystic ovaries is more prevalent among improved dairy breeds in the WCZ than in the cooler areas. A cystic condition may be due to failure of the pituitary gland to produce sufficient luteinizing hormone (LH) for ovulation or due to lack of receptors on follicles for LH. Cystic follicles can cause continuous estrous behavior (nymphomania), anestrus or estrous cycles approaching a normal condition. Some cows recover spontaneously while others can be restored by hormone therapy

Box 5.4. Merits and Limits of Some Antimicrobials to Treat Uterine Infections.

1. Penicillin
    * resistant bacteria may secrete penicillinase into the uterus protecting not only themselves but also susceptible bacteria
    * good choice for parenteral treatment when cow is systemically ill

2. Nitrofurazone
    * action inhibited by puss and blood
    * can irritate uterine lining
    * *Actinomyces pyogenes* is resistant
    * may adversely affect subsequent fertility

3. Aminoglycosides (mycins)
    * can inhibit protein synthesis
    * may reduce fertility (neomycins)

4. Sulfonamides
    * can be very effective
    * cannot be used in US and many other areas as sensitive tests will show residues in milk

5. Oxytetracycline
    * individual bacteria may be resistant but will not confer resistance to susceptible bacteria
    * effective for mixed infections
    * not effective if injected inter-muscular
    * good choice for early, mixed infections (intrauterine only)

Source: Adapted from Ferreira , 1991).

with gonadotrophins. However, if the condition reaches the stage where the tailhead is elevated or male characteristics have developed, the animal will not respond to treatment.

Anestrus and variable length estrous cycles can be frequent. How best to arrest these remains an unsolved problem of reproductive performance in the WCZ. Some recommend the use of gonadotrophins but this remains controversial because some cysts do not respond. Manual rupture of the cyst by palpation can cause bleeding which can lead to adhesions and prevention of an egg from entering the oviduct. Of those two choices, hormone injections are recommended. A third alternative is to use progesterone releasing intravaginal devices (PRID or CIDR) to mimic a luteal phase and reset the hormonal "clock."

Sometimes there is a problem in persistent CL which may have the opposite effect of cystic follicles in that the female fails to show estrus and if she does, fails to ovulate. A persistent CL may also denote some infection in the uterus. The condition may alleviate itself or respond to the use of prostaglandin treatment.

8. <u>Herd Size</u>

Under both intensive and extensive management systems, breeding efficiency tends to decline as herd numbers expand. It has been reported that the percentage of cows conceiving to first service declines at least 2% per 25 cow increase in herd size. Several causes have been indicated, such as reduced space per animal, crowding lessens the ability of an observer to detect estrus, floor surface conditions, and affects of social order in the herd. However, the most likely cause is that systematic management programs capable of dealing with the lactating herd or groups of heifers are not incorporated into the daily routines. Once programs of estrous detection and rebreeding are in place, a degree of focus on the individual cow becomes feasible.

Estrous Synchronization

Synchronization of estrus in heifers and cows can give herd owners the option of breeding at a predetermined time. However, success in synchronization depends on normal animals that are cycling. The process is not a remedial measure. There is caution in the use of any synchrony compound when nutrition is low as it will get little response and will add to the operating expense without benefit. More testing is needed in the WCZ before accepting transfer of this technology.

Currently there are several procedures being used among which are: an ear implant that contains the synthetic progestin, norgestomet and an injectable solution of norgestomet and synthetic estrogen, estradiol valerate. Another combines the use of oral progesterone (mel-engestrol acetate) and prostaglandin $F_2\alpha$. A third, and the only one labelled for lactating cows in the U.S., involves injection of prostaglandin $F_2\alpha$ (Hoard's Dairyman, 1990). However, in other areas of the world, the use of progesterone releasing devices (PRID and CIDR) are gaining popularity. Details on the use of any of these processes are available through veterinarians. Before attempting synchronization, a veterinarian or a representative of the AI organization should be consulted keeping in mind that synchronization should only be undertaken when cattle are cycling normally and the animals have sufficient body condition to support conception.

When successful, estrous synchronization requires some adjustments in herd management. For instance, 10 or more heifers entering the milking herd within a few days means that farm laborers must be prepared to observe heifers frequently to minimize problems at calving and for training heifers in the milking process. A third consideration is whether the calf facilities can handle the calves born from synchronized conceptions. Due to these pressures, most U.S. dairymen continue with the usual practices in the breedings of heifers under normal conditions.

The Male and AI

1. <u>The Male</u>

When bulls are used for breeding, their management to ensure libido and semen quality is important. Both libido and semen quality decline from heat stress with the greatest effect on the first stage of spermatogenesis with delayed effects on fertility. A young bull should reach at least 50% of expected mature weight before being used extensively. When bulls are run with the herd, particularly on pasture, it is best to use multiple sires in sequence and rotate these on a weekly basis otherwise breeding will be slowed because of the bull's needing time to rest and feed. Placing a bull with either heifers or cows on a dirt lot during the hours of darkness will often improve conception rate. When young bulls are to be placed in natural service, a prebreeding examination is recommended.

Under heat stress, bulls of *Bos taurus* breeds relax the cremaster muscles. This permits the scrotum to drop down from the body to serve as an aid in keeping the temperature of the testicles as low as possible. In older bulls, the scrotum will in some bulls drop down so low that the scrotum will be struck by the legs or even the hooves when walking.   This creates discomfort and decreased libido which can be factors in breeding on pasture.   Cautions on males are that they should be adequately fed to ensure libido and protected from severe heat stress in order to have acceptable quality semen.

2. Handling Semen

Throughout the WCZ, close attention to the handling of semen and AI techniques is needed. Consistently, minor mistakes are made in semen handling which often leads to lowering of sperm viability or non-viable sperm.   Another common shortfall is improper placement of the semen that can make it difficult for sperm to gain access to the site of fertilization in the oviduct. Recommended guidelines on storing and thawing of semen are in Box 5.5.

Fertilized Embryo Transfer (MOET)

Representatives of organizations producing fertilized embryos are travelling these days throughout the WCZ extolling the potential advantages of MOET (multiple ovulation and embryo transfer). Currently costs to produce an embryo transfer heifer calf in the US is about $200. The possibilities in gains are acceptable but MOET is not yet feasible in low economic situations. Genetic gain in cattle could be considerable but the cost per pregnancy will usually exceed $1,200 and per female coming into lactation 2.2 to 2.5 times this amount in much of

Box 5.5. Storing and Thawing Semen

1. Maintain minimum 15 cm of liquid nitrogen in storage tank.
2. Keep an accurate semen inventory near storage tank.
3. In removal of semen, keep tops of remaining straws below the frost-line. Use tweezers to remove straws.
4. Handle straws by tip and minimize light exposure.
5. Follow thaw temperatures and timing as directed by organization preparing the semen.
6. Warm water thawing is preferred to air thaw, if thawing temperature and time are adhered to.
7. Completely dry the straw, then load into the breeding gun in an area out of direct sunlight.
8. Inseminate cow within 15 minutes of thawing.

Source: Adapted from Ferreira, 1991.

the WCZ. Also, there is a slow return on the investment as the herd owner must wait four years or longer for results. Another misperception is in the performance of offspring from the use of MOET in local breed females. An embryo from parents in the temperate areas have the genes for high performance but when the embryo is placed in a small local cow, the expression of the genes for high milk yield will be constrained by the maternal environmental (in uterus) effects, e.g., the expected birth weight of a female does not reach the expected 40 to 45 kg, rather 22 to 25 kg, and the mature cow usually reaches only about 400 kg body weight. Because of higher risks from estrous synchronization in the WCZ the cost of MOET can go even higher.

To lower cost of MOET, flushing of selected cows is being replaced by the use of eggs taken from culled dairy cows at time of slaughter. Following this procedure the cost of MOET can be lowered.

The consensus is that MOET is not yet a sound economic approach for widespread herd reproduction of each generation.

Dairy operators whose entire or main business is selling milk will not obtain enough gain in milk from herd replacements to pay for MOET at this point in time. More needs to be done on the costs and benefits of various technologies, (AI, MOET, and sexing of semen) to determine the optimal responses under various conditions.

Some Indications

Reproductive rate is the major supportive trait for high animal biological efficiency as well as herd economic efficiency. Efficiency in reproduction in the WCZ can be satisfactory when: females are healthy; nutrition is sufficient to ensure cycling; females are at a steady body weight or gaining; and the females are attended with vigorous, fertile males 2 to 4 years of age or with estrous detection and AI using good semen and technique. If this be true, what is the problem? It is largely a lack of awareness. Low fertility is to a large degree, a temporary reaction to a negative energy balance, stress from climate and disease or parasites, any of which will cause temporary or permanent sterility. Also, behavioral responses need recognition; management of either the male and frozen semen is important. Caution must be exercised in the use of hormone therapy; synchronization of estrus requires certain criteria of animal performance for success. Time of day, floor surface and space per animal affect estrous detection, and reproductive efficiency can improve when operators of dairy enterprises decide to allocate proper attention.

Among helpful hints are: 1) memorize the heat cycle; pre-heat generally lasts about 6 hours; standing heat will last the next 6 to 18 hours; follicles will be released 24 hours or so after beginning standing heat; and the life of the eggs is for the following 6 to 10 hours, 2) prepare a written protocol so you as herd owner and employees have tangible material to follow, especially in detection for estrus, 3) heat detection must be followed with accurate records, 4) ensure semen thawing and inseminations are made properly, and 5) acceptable reproductive rate necessitates attention to details.

Breeds, such as Holstein, show a genotype by environmental interaction for reproductive performance in the WCZ. However, with adequate feeding and close attention to heat detection much of the effects can be overcome.

Conception to first service is in the range of 40 to 60% in the temperate areas but often is <40% in the WCZ which is another reason for attention to individual cows.

The relation between high milk yield and reproductive traits is associated phenotypically and genetically with reduced reproductive performance in lactating cows. The rate of breeding efficiency is compromised through delays in ovarian function and lower conception rate by the demands of energy for high milk yield. Nevertheless, management decisions to obtain efficient reproductive performance have considerable impact (Nebel and McGillard, 1993).

CHAPTER VI

HERD HEALTH PROGRAM

Background

Investment in commercial dairying will not be economically viable unless accompanied by well executed health programs. Although effective treatment of health problems is critical, the prevention of diseases has the greatest economic benefit. Herd management practices in cow handling, nutrition, milking procedures, sanitation, and housing play major roles in predisposing the individual animal as well as herds to diseases. Another important reason for minimizing animal diseases is that some are transmissible from cattle to humans. Of over 200 communicable diseases, near one-half are threats to humans and about 80 are transmitted naturally between vertebrate animals and humans. Some of the more common are undulant fever, tuberculosis, cow pox, ringworm, rabies, and anthrax. In addition humans are susceptible to many of the same parasites, e.g., roundworms, tapeworms and flukes.

Control of diseases which are spread through vectors, wild animals or by humans should be the responsibility of governments and international agencies, but control of other diseases and endo-parasites is the responsibility of herd owners. Vaccinations may be required by national law or be voluntary. However, unless nutritional levels and farm sanitation are adequate, the effectiveness of vaccination can be low or nil. Veterinarians can recommend removal of animals from herds which are obviously ill, but herd operators are the key decision makers. Removal of animals which are carriers and shedders of disease organisms can reduce health cost by near 50% and would have a comparable effect on total animal production. A calf having pneumonia or infectious scours has three times greater chance of serious health problems as an adult than a normal calf. Serious problems as a calf will lead to a cow that falls below herd average in milk yield and rate of breeding. The risk of a health malady occurring which will significantly influence milk yield is about five times greater for cows seven years or older than for cows in first lactation unless there is a good health program. These risks mean that rearing of herd replacements for 15 to 20% of the herd per year can be needed to remove "high risk" cows. As attempts are made to increase level of herd numbers and production, careful management is required to keep health disorders, such as mastitis, from becoming more frequent.

In countries of the WCZ, endemic diseases like rinderpest and aftosa are generally deemed as major concerns, but in herds of cattle and buffaloes with recordings, the most frequent causes for termination of lactation and involuntary culling are similar to those of herds in cool climates, mainly "nuisance problems," e.g., foot rot. Collectively, it is clear that decisions on the part of herd managers are the main means of reducing health problems. Thus herd management practices combined with a veterinary program can be most effective in optimizing production and profitability.

Many countries in the WCZ have declared policies to attain and/or maintain freedom from diseases, the eradication of economically important infectious diseases, provide diagnostic services, and give free veterinary services. There are usually at least three major fallacies in free services: 1) government resources are so limited that both services and therapeutic drugs are too restrictive for needs, 2) about 80% of public investment in promoting livestock development will go to support health services leaving little for research and other needed farmer assistance, such as monitoring animal performance on farms for gathering data to serve as technical inputs for the national trends, and 3) the virulence of diseases can change too rapidly for the best effective responses through usual government channels. To illustrate, several countries of the WCZ report serious concerns over the incidence of mastitis. Proposed action through government is to develop resources to produce a vaccine for use nationally. Two risk factors which could lead to low benefits are: a) dairy operators would use a vaccine as "false security," meaning low attention to milking sanitation thereby making a greater threat of mastitis; and b) no country has yet developed a multi-valent vaccine which holds against mastitis for a significant period of time. The recommendation to dairymen is to exercise careful herd management and form a linkage with an experienced veterinarian.

The major thrust of this chapter is to show that knowledge of the physical elements of the environment and the interactions with the physiological functions of cattle permit decision making by herd operators on management practices to ensure high efficiency and maintenance of good health.

Predisposition to Diseases

Health problems arise principally from environmental stresses. An important result of stress is that animals have a depressed immune response due to immunosuppressive effects of an increased secretion of corticosteroid hormones. A depressed immune response results in increased susceptibility to infectious diseases, susceptibility to normal antivirulent bacteria, activation of latent viruses, and poor responses to vaccines. Effect in the WCZ is confounded by malnutrition which as well as being a stressor can impair immunoglobin production, especially when protein level in the ration is deficient.

Box 6.1 enumerates some of the most common predispositions. Of these, nutrition deficiencies are most significant.

Box 6.1. Predisposition to Poor Animal Health.

- Physical injury
- Thermal Stress
- Genetic Abnormalities
- Nutritional deficiencies
- Pathogenic organisms

Viruses	Bacteria
Fungi	Protozoa
Endoparasites	Ectoparasites

- Age of animal (very young or old)
- Previous exposure
- Failure to receive colostrum at birth
- Failure of intra-uterine transfer of antibodies

Although there is some evidence of direct genetic resistance to certain health maladies, immunity to a disease process may be attained to an appreciable extent in certain cattle through contacts with limited amounts of infection. Contact may be either natural, e.g., through feeding, or artificial, e.g., through vaccination. Much of the resistance shown by some cattle to certain diseases is probably in the intrauterine environment through passage of antibodies to the fetus and through colostrum feeding postpartum. Such seems to explain at least short term resistance to anaplasmosis, vesicular stomatis, East Coast fever and to a lesser extent foot and mouth disease. The characteristic skin and pelage of cattle can be important indirectly for susceptibility to the invasion of disease organisms, e.g., ticks have more difficulty penetrating the skin of Zebu cattle than European dairy breeds.

Another factor which can play a role in resistance is the ability of animals to develop their immunological defense mechanisms at an early age. In cattle, there may be differences in activity of the thymus gland which plays an important part in building defenses.

Age is also a factor in predisposition with the very young (< 6 mo) and old (>7 yr) having the least tolerance. Cattle show the greatest resistance between 1 and 4 years, with the period between 1 and 2 years being best. This has been clearly demonstrated in the transfer of cattle from cool climates to the WCZ. It is recommended that cattle be under 2 years of age when introduced into an anaplasmosis area and that the introduction be made in the season when vectors are lowest.

Infections

The most common viral infections in the WCZ are listed in Box 6.2. At present, only bovine malignant catarrhal fever, rinderpest, rift valley fever and foot and mouth disease can be successfully controlled through vaccination but there is limited success for rabies (Box 6.3). For animals suspected of exposure, immediate actions should be taken for isolation to exclude contact followed by disinfection or sanitizing the area. Other methods are identified in Box 6.4.

Box 6.2. Common Viral Infections.

- Blue tongue
- Bovine malignant catarrhal fever
- Ephemeral fever
- Calf Diarrhea
- Foot and mouth (Aftosa)
- Lumpy skin disease
- Rabies
- Rift valley fever
- Rinderpest
- Vesicular stomatitis (sore mouth)

Bacterial Infections

Infections from both bacteria and protozoa are widely spread throughout the WCZ, e.g., anthrax and brucellosis (Box 6.5). As for the case with viral infections, only a portion of the bacterial diseases and none of the protozoal infections can as yet be successfully controlled by use of vaccines (Box 6.3), thus actions by herd managers are quite important in control and limitations in herd losses (Box 6.4), in particular for bacterial infections. For protozoal diseases, isolation and separation are recommended followed by concentration on means to control the vectors.

Protozoal Infections

Protozoal diseases like East Coast Fever and trypanosomiasis are acclaimed as causes for major losses of animals. But coccidiosis generally causes higher morbidity losses and mortality among all domestic animals and fowl. It is frequently a cause of high mortality in calves up to one year of age in the WCZ with greatest occurrence during conditions of high humidity (rainy season).

It is common practice to use antibiotics, either injected or fed, to aid in control and reduction of virulence of bacterial, viral and protozoal infections. However, the value of prophylaxis is best when sanitation of the facilities is good and wetness around the premises a minimum.

Box 6.3. Vaccinations in Animal Health.

<u>Successful</u>

Anthrax
Blackleg
Bovine pneumonic pasteurellosis (shipping fever)
Bovine malignant catarrhal fever
Brucellosis
CBPP (Contagious bovine pleuropneumonia)
Foot and mouth disease (Aftosa)
Heartwater
Rift valley fever
Rinderpest
Tetanus

<u>Limited Success</u>

Coccidiosis
Leptospirosis
Mastitis
Rabies

<u>None Available</u>

Dermatophilus congolensis
East coast fever (Theileriasis of cattle)
Trypanosomiasis
Tuberculosis
Vesicular stomatitis

Internal Parasites

There are over 300 kinds of internal parasites of economic significance in the U.S. and up to 10 times this number for the world. Parasitism exists primarily subclinically resulting in unnoticed economic losses. Essentially, parasite infestation is a herd disease, hence measures for control are effective only if applied to the whole herd. Eradication of most infestations is unfeasible in the WCZ, if not impossible and perhaps undesirable. Sanitation along with selective, timely medication appears the most effective measure for control.

Box 6.4. Disease Control Measures.

<u>Viral infections</u>

- Isolation or exclusion of contact
- Slaughter
- Disinfection
- Vaccination
- Acquired immunity
- Prophylactic feeding with antibiotics

<u>Bacterial infections</u>

- Isolation or exclusion of contact
- Slaughter
- Disinfection
- Vaccination
- Acquired immunity
- Prophylactic feeding with antibiotics
- Control vectors

<u>Protozoal diseases</u>

- Control vectors

<u>Fungal diseases</u>

- Isolation or exclusion of contact

The common worm parasites are broadly classified into roundworms, flukes or trematodes, and tapeworms or cestodes (Box 6.6). Some worms cross the fetal membrane, therefore calves are infested at birth. Grazing animals become infected with roundworms through contaminated pastures. After ingestion, the female roundworms deposit eggs in the digestive tract; which are then passed in the feces. Unless the droppings are exposed to direct sunlight for rapid drying, the eggs hatch into free-living larvae ready for another host and cycle. Tapeworms behave in much the same fashion. Dairy operators with Holsteins often keep newborn calves on slatted floors, up to about 6 months of age, essentially a worm free environment. This means they are highly susceptible, so when turned on

Box 6.5. Bacterial and Protozoal Infections Common in the WCZ.

<u>Bacterial</u>

- Actinomycosis
- Anaplasmosis
- Anthrax
- Blackleg (Black-quarter)
- Brucellosis (Contagious abortion)
- Contagious bovine pleuropneumonia
- Foot rot
- Heartwater
- Hemorrhagic septicemia
- Infectious conjunctivitis
- Leptospirosis
- Mastitis
- Pneumonia
- Tetanus
- Tuberculosis
- Vibriosis

<u>Protozoal</u>

- Babesiasis (Redwater, Piroplasmosis)
- Coccidiosis
- East Coast Fever
- Trichomoniasis
- Trypanosomiasis

pasture at 6 to 8 months, they rapidly lose "their bloom". The hair coat becomes shaggy and they lose weight. Supplementary feeding helps until the calves develop some immunity or

Box 6.6. Major Internal Parasites of Cattle in WCZ.

Parasite	Symptoms	Prevention
Flat or tape worm (*Monezia expansa*)	Unthriftiness, weight loss, invades small intestine	Reduce mites (intermediate host)
Hookworm (*Bunostomum phleobotomum*)	Anemia, rapid weight loss, possible intermittent constipation and diarrhea	Avoid wet pastures and pens, good sanitation practices
Lungworm (*Dictyocaulus viviparous*)	Respiratory difficulties, labored breathing, elevated temperature, lungs are affected	Strict sanitation and deworming routines, avoid spreading infected manure
Liver fluke (*Fasciola hepatica*)	Anemia, digestive disturbances, unthriftiness, liver tissue destroyed, severe liver damage in acute cases	Prevent contact of cattle with snails, avoid wet pasture
Roundworms (*Onchocerca spp.*)	Diarrhea in calves, anorexia and loss of weight, most commonly located in brisket	Difficult to prevent
Stomach Worms Large	Anemia, loose weight, digestive disturbance	Avoid overcrowding, rotate pastures
Medium	Profuse watery diarrhea, weight loss, abnormal thirst	Avoid overcrowding
Small or Bankrupt worm	Unthriftiness, watery diarrhea, digestive disturbance	Proper pasture management
Threadworm (*Strongyloides papillosus*)	Intermittent diarrhea, bloody feces, lesion on foot like foot rot	Keep facilities and pasture dry

Source: Adapted from Yakstis, 1983.

resistance. There is evidence that immunity can be developed. If calves or heifers are not exposed to a large number of larvae, they can become partially immunized and have fewer worms later. After 2 or 3 months, infestation in calves may increase rapidly up to about 9 months of age, then more slowly thereafter with the incidence tending to decrease at about 18 months of age.

Liver flukes can be a large problem in lowland areas of the WCZ, e.g., in Puerto Rico about 75% of all livers of cattle processed in slaughter houses from herds in the wetter areas are condemned. Chemical treatment can be effective, but fencing off or drainage of swampy areas serves as the best control.

Although worms are capable of killing cattle, their effects on the hosts are difficult to assess. By the time the secondary effects become evident, e.g., shaggy hair coat, drooping ears, extension of the neck and head when standing, or extreme emaciation, the chances of the cattle returning to normal performance are slim. If the shaggy coat stage is reached, puberty is delayed or does not occur. Should conception eventually occur, milk yield will be about 30% below herd average and rebreeding slow.

It is a fair assumption that almost universally, parasitism is the normal state. But if the levels are not excessive and the cattle not malnourished, their relationship can be more beneficial than pathogenic. This is due to the factors of resistance and immunity. Coupled with this is the fact that very few parasites live entirely and replicate within their host. Their life cycle is associated with sexual reproduction and several stages of development which require different environments, hence the level of parasites within a host is directly related to level of transmission. The rate of transmission can be regulated largely through management methods based on an understanding of the life cycles of parasites.

Some conclusions are that parasites rob the animal of nutrients, do harm to vital organs and increase susceptibility to infection by bacteria or other disease producing agents. The migration of developing parasitic worms through organs and tissues interferes with normal body functions resulting in poor utilization of pastures, labor and space. The extent of worm infestation rests largely with the herd or farm manager as control is largely breaking the life cycle of parasites. High emphasis is given to control measures for internal parasites since the author has viewed more herd failures resulting from infestations throughout the WCZ then any other cause. Imported bred heifers are highly susceptible with losses in the first year, up to 30%, attributable directly or indirectly to internal parasites. A large portion of the debilitation was caused by low attention to wetness of pastures or using cut forages from low lying areas and poor sanitary conditions around barns or shelters.

For control of internal parasites it is not possible to outline a simple program. Each farm needs to tailor their control program to their environment and management. The key issues are to increase resistance and lower challenge. Resistance should rise from colostrum feeding and appropriate nutrition. Challenge can be reduced through sanitation, use of deworming medications and reducing the stress on the cattle whenever possible. It is cautioned that lactating cows should only be treated with products labelled for same.

External Parasites

External or arthropod parasites impair efficiency of cattle as a nuisance. Some serve as transmitters of disease and decrease the quality of products (Box 6.7). There are several thousand species but less than 100 are considered hindrances to livestock. Economic losses from arthropods (invertebrate animals, such as insects, arachnids and crustaceans) are usually highest in the WCZ, more from morbidity than mortality, unless the parasite transmits a disease. Morbidity losses are principally from reduced feed intake resulting in reduced rate of gain, e.g., cattle treated to control hornflies gained 5 to 20 kg more per month than non-treated animals.

The need for protecting cattle against arthropods is recognized even under primitive conditions, e.g., pastoral herders in Africa use smoke to curtail annoyance from flies; Indian villagers keep their dung cake fires going well into the night to keep mosquitoes away, during the monsoon season; and children are assigned in numerous countries to remove ticks. Blood sucking parasites are most numerous, such as mosquitoes, gnats, flies, lice and ticks.

The annoyance of mosquitoes and flies cause cattle to congregate and use proximity of each other in freeing themselves of the irritation of insects thereby reducing feeding time. Some parasites may produce toxins which are injected into the host during the process of piercing the skin to suck blood. These may create severe local skin reactions particularly evident in white haired Holsteins.

Herd owners seldom recognize their losses when insects become highly irritant to lactating cows, thus when laborers come from homes without insect control, they must be consistently reminded of planned control measures. Sprays or powders for controlling insects have cost but their systematic use yields quite high returns in cow performance.

Larvae of cattle grubs (*Dermatobia spp.*) cause damages to cattle hides. They are best controlled by dusting an insecticide on the back of cattle near the time they are expected to emerge to break the life cycle. Treatment with systemics about 4 to 5 months before emergence is expected is the common practice in the U.S. Treatment may consist of pour-on, injectable insecticides, as well as dusts.

Screwworms are a wide problem in Latin America and Africa. Eggs are laid by the blow fly into wounds or around the naval of neonatal calves. The eggs produce larvae which will destroy tissue and consume blood and lymph until the animal dies. When screwworm is identified, attention at calving is a good practice because dipping of the naval cord in an iodine solution quickly dries up the cord.

Midges (*Culicoides spp.*) transmit the virus for blue tongue. They are extremely small flies, breed in low lying wet areas, and mainly affect sheep, although they attack cattle too.

There are many types of mosquitoes. Some transmit Rift-Valley fever to sheep and cattle and in areas of the WCZ have been identified as a vector for anaplasmosis, not by passing the organisms through their body, but by mechanical transfer of blood from one animal to another.

The ability of ticks to transmit diseases to their offspring by transovarian transmission is the reason they can be highly detrimental. With this capability, there is no need for each generation to pick up a disease from infected hosts. There are many species of ticks but eight are major disease transmitters. The blue tick (*Boophilus spp.*) is a carrier of Redwater and anaplasmosis; the red tick (*Rhipicephalus evertsi*) transmit Redwater, East Coast fever and anaplasmosis; and the brown tick (*Rhipicephalus appendiculatus*) is a carrier of East Coast fever, anaplasmosis and Redwater.

Tick control consists of periodic dipping or spraying with insecticides but several insecticides must be used as ticks quickly build up resistance. Ticks with more than one host are more difficult to control because they spend only part of their time on domestic animals and exist in reserve numbers on uncontrolled secondary hosts. Interval for dipping or spraying is recommended at 14 days for one host ticks, 7 days for two host ticks and 5 to 7 days for three host ticks.

By using long pasture rotations, operators destock long enough so that most of the immature ticks die without finding a host. The rest period must be 3 to 4 months to be effective, however. The usual 2 to 3-week rotational grazing pattern contributes little to tick control. Cattle should of course be resprayed or redipped before returning to a rested pasture.

The vampire bat (*Desmodus nufus*) is not an arthropod but feeds on the blood sera of animals and humans and transmits rabies. In some areas of Latin America, cattle cannot be left in the open at night because bats will attack while the cows are resting. Dairy breeds appear more susceptible than Zebu or local Criollo.

Photosensitization is a health disorder generally associated with exposure to sunlight, but porphyrinemia (porphyins in the blood) and toxins from plants and biting insects can also cause photosensitization. Cattle with light pigmented skin are most susceptible. Dairy breeds with non-pigmented skin are most vulnerable. If recognized in time, the condition may be arrested by sheltering but once the condition occurs, it will repeat when cattle are again exposed to sunlight for extended periods.

Non-Infectious Health Disorders

Some health disorders characterized as non-infectious but often add to risks in dairy herds are; bloat, ketosis, milk fever, grass staggers, fluorine poisoning, iodine deficiency, phosphorus deficiency, and selenium poisoning. Bloat and grass staggers are rather common on pasture grazing in cool climates but these disorders have not been reported on tropical grasses. Both milk fever and ketosis are definitely on the rise in commercial dairy herds of the WCZ because of long dry periods, nutritional imbalances and low levels of chewable fiber when fed fresh cut, low DM grasses. At this time, there are far too many cows affected by laminitis due to low pH and high rumen acidity which leads to high ketone body production. This problem was addressed in Chapter III (Nutrition). The incidence of milk fever may rise from either low phosphorus or excess calcium. Both impair the calcium to phosphorus ratio.

Excess selenium has been identified in feeds of certain areas, especially when irrigation is used, such as in the Punjab state of India, but testing has not been widely employed. Thus far fluorine toxicity has appeared in only a few areas, mainly low lying coastal. Frequently, the condition pica (perverted or depraved appetite) is associated with low phosphorus in tropical grasses. Lashing out of the tongue and the chewing of wood are the usual symptoms.

Problems in Intensive Enterprises

Small space per head in concentrated dairy operations generally leads to increasing sanitation problems particularly in feeding and resting areas. Close confinement tends to foster health disorders such as foot rot, mastitis, diarrhea, coccidiosis, vesicular stomatitis, foot and mouth, leptospirosis, mycobacterial skin lesions, external parasites and intestinal disorders. High standards of sanitation are required at all times to prevent outbreaks of some of these health disorders from going rampant through the herd. One of the greatest weaknesses in large dairies is allowing manure buildup in dirt lots provided for resting and for observation of heats.

When young calves are penned together, consistent attention is needed to prevent rapid spread of diseases. This is the main basis for recommending that young calves be penned individually for at least the first month. Meticulous sanitation, adequate nutrition and perhaps disease prophylaxis are musts to minimize health problems among artificially reared calves.

Calf diarrhea or scours is often a serious problem in herds of improved breeds in the WCZ. It can be caused by a group of infectious diseases which share many clinical features (Mechor, 1993). Two basic mechanisms are responsible, hypersecretion and malabsorption. *E. coli* bacteria is the most common cause of the hypersecretion form and is commonly seen in calves during the first 3 to 5 days of life. The condition can appear suddenly and be severe. An endotoxin produced by the bacteria causes exaggerated intestinal secretion resulting in serious dehydration. Viral diarrhea pathogens, such as rota and corona virus, usually appears between 5 days and 2 to 3 weeks of age. These cause damage to the lining of the intestine which leads to impaired absorption, a malabsorptive diarrhea. Infected calves will show diarrhea until the damaged intestine heals.

The major abnormalities identified with diarrhea are dehydration, acidosis, electrolyte abnormalities, and negative energy balance or hypoglycemia. Herd operators should recognize that diarrhea in neonatal calves cannot be effectively treated with antibiotics alone. Electrolyte therapy should be the central of treatment. Oral antibiotics are generally over used as a treatment. In fact oral antibiotics can induce diarrhea through elimination of normal beneficial bacteria in the intestine and also produce drug resistant strains of bacteria.

The goal is not to have to treat numerous cases of diarrhea. When incidence of diarrhea rises, investigation into the management system for cause(s) should be pursued. For example, diarrhea after 14 to 18 days of age suggests problems with coccidiosis.

Coccidiosis can be another serious "nuisance" with cattle at all ages. Almost all domestic mammals in the WCZ carry the causative protozoan. In close confinement, losses occur in calves by a build-up of infection which is transmitted from adults. *Eimeria zurnic* and *E. bovis* are the species most often associated with clinical cases. Coccidiosis is most common in calves from 1 or 2 months of age to 1 year and is usually sporadic during the wet season but may occur at any time. Similar to malabsorptive diarrhea, the pathogenic coccidia can cause damage to the mucosa of the lower intestine, cecum and colon. In light infections, the most characteristic sign is watery feces and the animals show indisposition lasting a few days.

Sulfonamides continue the drugs of choice for treatment. Although clinical infections are self-limiting and subside spontaneously in about a week, prompt medication usually results in shortening the course of infection. Sulfaguanidine or the readily absorbed sulfonamides, such as sulfamerazine can be used. Some herd owners in the temperature areas use prophylaxis compounds like liquid amprolium in the milk fed to calves and in milk replacers. The key to keep in mind is that coccidiosis control is second only to good nutrition for heifer growth.

Sanitation

Sanitation has been stressed as a means of minimizing health problems. Ultimately, it is the most effective method of controlling diseases. But few people really know what constitutes good sanitation. For sanitation to be effective in dairy operations, one should be aware of appropriate cleaning procedures, why it is important and how it can be accomplished effectively. Herd owners should prepare a series of protocols on preservation of premises and cleaning, e.g., a profile for calf facilities, the comfort stalls, the exercise lot, the milking machine, etc. Udder and teat sanitation must be of primary consideration as will be shown later (Chapter VII).

Responsibility for execution of each protocol should be recorded for all to see. Where hired labor is employed, the herd manager should follow with daily inspections. Particular care should be taken to see that instructions are understood when languages differ.

The most reliable means of destroying disease organisms and parasites at certain stages is cleaning and disinfecting. However, the latter will have little effect unless the surface is first cleaned. The best cleanser and disinfectant depend upon the type of surface. Lye, caustic soda, cresol and chlorine based products are successful cleansers, along with liberal applications of water and detergents. No matter how primitive the situation, there are usually some means of maintaining reasonably good sanitary standards. Compounds having broad availability are lime and some form of oil products, such as kerosene or diesel fuel. Sunshine is another very effective agent, and even changing the soil, such as in the exercise lots can be quite helpful. It should also be kept in mind that good sanitation practices enhance the value of therapeutic agents. The Merck Veterinary Manual and similar publications are good resources for recommendations on sanitizing, therapeutic agents, treatment procedures and identification of disease conditions. The book, "Diseases and Parasites of Livestock in the Tropics" (Hall, 1977) provides a good overview of problems likely most prevalent and is particularly good on symptoms in layman's terms.

An effective herd health program is time consuming and costly but essential. Electronic devices are under test which look promising to supplement health programs, e.g., there is an ankle bracelet being marketed. A small antenna tied into a computer monitors individual cows. The cow's activity is measured by the number of steps she takes. When steps increase significantly while milk yield declines, the cow is likely coming into estrus. But if her activity drops along with milk the cow could be getting sick. Changes in activity can identify the abnormal cows thereby reducing the number of cows needed to be viewed closely. The recommended parameter for a cow in heat is 100% increase in activity from previous day when the cow remains in the same group.

Another device is to utilize an electronic meter used to measure milk yield and electrical conductivity of the milk. The meter can also tell how long each cow takes to complete milking. A computer can be programmed to provide the average milk yield for the past 10 days. A viewer then scans the herd looking for significant variables from the cow's norm. The abnorms can be used in the same fashion as for the pedometer in detection of estrus and possible illnesses. The manufacturers claim that changes in electrical conductivity of the milk can be used for identifying cows in the early stages of mastitis, e.g., 24 to 48 hours earlier than the showing of flakes on a strip cup. Data on milking speed can be used for grouping of slow milking cows rather than having them intermixed which can be a factor of milking in large herds. The newer methods of monitoring cows within the herd have potential for adoption as the technology becomes less expensive compared to other inputs.

New devices presently available can serve as aids in health programs but these should not lessen responsibilities for the basics of herd health that of building resistance in the animals and minimizing challenges. Resistance is increased commencing with colostrum feeding followed by vaccination and good nutrition. These inputs improve our power to lower the challenges from viruses, bacteria, protozoa and parasites as disease occurs whenever the challenge is able to overcome resistance. Challenge can not be eliminated nor can complete resistance be developed but herd managers must strive to raise resistance and lower challenge (LaDue, 1993).

CHAPTER VII

MILKING MANAGEMENT AND
MASTITIS CONTROL

Background

Topics in the heading are discussed separately from the health program since they constitute the greatest economic loss from milk production and premature culling. In the U.S., about 70% of the economic loss in milk is due to unseen subclinical mastitis, another 14% is due to culling and death loss, while 8% is from treatment costs. Some contend that mastitis is a "western" problem with high producing cows and use the supposition of low frequency in indigenous cattle as a factor of preference for their use. However, in studies on local breeds, the incidence of mastitis is on average lower but so is milk yield. On an adjusted milk yield basis, there is no clear evidence of greater resistance in local breeds (Talbott et al., 1992). A recent survey of "city dairies," using mainly buffaloes, within the city of Karachi, Pakistan, the level of somatic cells in the milk was high. There were 78% of the samplings with counts > 400,000 and of these 68% were > 1 million. The high cell counts indicate high levels of subclinical and perhaps clinical mastitis.   The conclusions are that mastitis is a global problem and we have as yet little evidence of significant genetic resistance. However, U.S. AI sires are being ranked on the somatic cell count scores (SCCS) of their progeny as estimates of possible genetic differences. Near all major dairying countries are investigating the value of progeny testing of sires for SCCS as a means of improving resistance to infections leading to mastitis.

Dr. Nelson Philpot, a leading authority on these topics, has kindly consented to paraphrasing from his many writings especially his report on: "Milking Procedures and Management Practices for Improving Milk Quality" (Philpot, 1991).

A number of countries have begun to initiate penalties on deliveries of milk with high bacteria counts, e.g., in the U.S. as

- 124 -

of July 1, 1993, producers cannot legally send milk to processors that has >750,000 SCC per milliliter. The milk market in Ontario, Canada drops the price on milk with >750,000 SCC by $0.35 per hundred weight at first violation; a second violation doubles the drop ($0.70) and a third offense deletes $1.05 per hundred weight. Puerto Rico and Taiwan have also initiated penalties based on bacteria counts per milliliter of milk to protect consumers and growth in the markets for milk products. Cows with SCCs exceeding 400,000 eat the same amount of feed as low cell count cows but produce 17 to 25% less milk. All producers must recognize that sanitary milking practices and mastitis control programs are essential to profitable dairying (Philpot, 1991).

Milking Procedures

Recommended milking procedures are described briefly in Box 7.1 with additional comments as follows:

1. <u>The Cow's Environment</u>

The environment should be as sanitary as practical; milking time should be a consistent routine; and the cows should not be frightened or excited because stress causes release of hormones into the blood stream that interfere with milk letdown. Unsatisfactory milk release is predisposing to mastitis.

Udders may need on occasion to be clipped or the hair singed, as short hair reduces contamination of the milk with dirt, manure or bedding. Short hair also eases cleaning and drying of the udder prior to milking.

2. <u>Check Foremilk and Udder for Mastitis</u>

Clinical mastitis can be detected by milkers using hands to physically examine the udder and by using a strip cup or plate, such as the California Mastitis Test, to examine foremilk prior to each milking. Such procedures aid in detecting hot, hard and enlarged quarters as well as clotty, stringy or watery milk.

Forestripping of milk directly onto the floor can be used but should be followed by immediate flushing of the floor surface. Never strip directly into the hand as this will spread organisms from teat to teat and cow to cow.

3. Teats and Surface of Udder

Proper washing and massaging the udder sends signals to the pituitary gland of the base of the brain which, leads to release of oxytocin needed for good milk letdown. Cows that are frightened have a release of the hormone adrenalin that will interfere with the action of oxytocin. The goal is to milk clean and dry teats, thus all teats, as well as the lower udder, should be washed and dried. Wet udders and teats will spread mastitis and cause high bacteria counts in herd milk.

In milking parlors a low pressure water hose should be used to deliver sanitizing solution to the teats and udder. Wetting the entire udder should be avoided as it makes it difficult and time consuming to get the entire udder dried. Paper towels are recommended to finish drying the udder and teats.

Box 7.1. Recommended Milking Procedures.

1. Provide cows with clean, stress free environment.
2. Check foremilk and udder for mastitis.
3. Wash teats and ventral surface of the udder.
4. Use premilking teat dip (optional).
5. Dry teats thoroughly.
6. Attach teat cups within one minute.
7. Adjust milking units as necessary.
8. Shut off vacuum before removing teat cups.
9. Dip teats with an effective teat dip.
10. Disinfect teat cups between cows (optional).

Source: Philpot, 1991.

---

4. Premilking Teat Dip

Although disinfecting teats with a germicidal dip before milking is classed as optional, this procedure is recommended as a means of mastitis control in the WCZ. Predipping can reduce mastitis caused by environmental organisms by about 50%. It is imperative that teats disinfected be thoroughly dried before attaching teat cups to avoid contaminating the milk with germicide residues. Recommended procedures are: clean teats, forestrip, predip teats and allow recommended contact time, of 20 to 30 seconds, dry teats to remove germicide, then attach teat cups.

5. Drying Teats

Irrespective of the method used to clean the teats and lower udder, these surfaces must be dried before milking. Failure to fully dry the skin surfaces results in water laden with microorganisms draining down and being drawn into teat cups; a sure way to increase microorganisms in milk and possibly lead to mastitis. In herds of the WCZ, it is common to wet the cows all over prior to milking for some cleaning but the main intent is an aid in cooling. If practiced, the procedure must be done far enough in advance of milking for the cows to drip dry otherwise the hazards of mastitis become high.

6. Timing for Teat Cup Attachment

Attach the teat cups as soon as the teats are obviously enlarged from milk letdown; usually within 1 minute after initiating udder preparation. The procedure must be executed carefully to prevent entrance of excessive air into the milking system; a common cause of vacuum fluctuation on the whole system.

Maximum internal udder pressure commences about 1 minute after preparation is begun and endures around 5 minutes. Since the majority of cows of improved dairy breeds milk out in 3 to 5 minutes, the timing of the teat cup application is important in maximizing the action of the hormone oxytocin in total milk extraction.

7. Adjusting the Milking Units

The milking units need observation while attached to ensure correct adjustment and to prevent liner slips. Teat cups seated high on the teats cause irritation to the teat and can contribute to mastitis. Inappropriately aligned units reduce milk flow and increase strippings. Slipping or squawking teat cups contribute to more machine induced infections than any other factor. If slips on the liner occur at the same time as the liner opens, droplets of milk will be propelled against the end of the teat at high velocity. Should the droplet contain microorganisms, they may penetrate into the teat canal and cause mastitis.

8. Vacuum Control

Vacuum to the teat cups should always be shut off before removal of teat cups as risk of infection rises if the cups are removed under vacuum. A minute or more of overmilking does not usually pose threats of mastitis but the risk of infection is greatest during overmilking.

9. Post Teat Dipping

Dipping of teats immediately after milking using an effective germicide is perhaps the most important step to reduce the incidence of new infections. The entire teat should be dipped. Most commercially available dips will reduce new infections by at least 50%. Some cautions are: never pour unused dip back into the original container, and clean cups well before milking. The tissue surrounding the teat canal, mainly the sphincter muscle which keeps the canal closed between milkings, is the first barrier against organisms causing mastitis. The sphincter remains partially dilated for up to 2 hours postmilking. Directing cows leaving the milking operation to fresh feed is strongly recommended as a stimulus to keep them standing until the sphincter muscle tightens and reduces the size of the teat opening.

Keratin in the teat canal can partially block the teat opening. This aids in defense against microorganisms, therefore efforts should be made to maintain the natural health of the teat opening (Philpot and Nickerson, 1991). It is cautioned that the feed offered to cows after milking be fresh to attract the cows since in large herds cows will have been standing a long while awaiting milking and become anxious to lie down as soon as they can.

10. Disinfect Teat Cups

This must be classified as an optional procedure. Cup liners frequently become contaminated with mastitis causing organisms because of intermixing of infected and uninfected cows in the order of milking. The most common method of control is disinfecting cup liners between cows through dipping them in disinfectant solution for a few seconds. The caution is that unless done properly, it may spread mastitis rather than reduce the incidence. This arises because when the cups are dipped milk is rinsed into the disinfectant solution, if over used, the solution may come to look like skim milk.

Milking procedure management serves to stimulate milk letdown, reduce microbial contamination of milk, decrease the need for stripping, decrease milking time, reduce the spread of udder and teat infections, and increase profits.

Microorganisms Causing Mastitis

Microorganisms that most frequently cause mastitis are usually grouped as: contagious, environmental, opportunistic and other. The contagious group are in the udders of infected cows and spread from cow to cow during the milking process. The most important of the group are *Staphylococcus aureus* and *Streptococcus agalactiae*.

The major sources of the environmental organisms are the cows' surroundings. Most of this group of organisms gain access to the interior of the udder between milking or during the dry period. Opportunistic microorganisms are normally found on the surface of the udder and teats, therefore are a constant threat of infection. Other less common microorganisms may lead to mastitis, such as yeast.

Control of Mastitis

Of most importance in planning a control program is recognition that mastitis is an infectious disease and that all methods of dairying provide suitable conditions for the spread of organisms which can cause mastitis. Mastitis differs from diseases like brucellosis and tuberculosis since the main goal cannot be eradication except for *Streptococcus agalactiae* mastitis (Philpot, 1991). The herd is the unit of control. The level of infection in a herd is independent of the level of infection in adjoining herds. This means, as Philpot (1991) states: "The primary need is not for a herd program applied nationally."

A control program ought to increase economic returns, reduce occurrence of new infections, show evidence that clinical cases can be reduced and the program should be flexible for easy modification as new knowledge becomes available from research. Features of a comprehensive plan of control in herds with emphasis on reduction of risks from contagious and opportunistic microorganisms are listed in Box 7.2. Needs to minimize infections from environmental microorganisms, primarily the coliforms are enumerated in Box 7.3.

The effectiveness of a control program can be determined as the level of infection (percent of cows with infected quarters). Level of infection depends upon the rate of new infections and duration of infections.

While mastitis occurs at different levels of severity (Box 7.4), clinical and subclinical infections are more common than the other four forms. Usually for each clinical case in a herd there

---

Box 7.2. Comprehensive Plan for Mastitis Control.

- Proper milking hygiene
- Use of functionally adequate milking machines
- Dipping teats after milking
- Treat all quarters of all cows at drying off
- Prompt and adequate treatment of all clinical cases
- Culling of chronically infected cows

Source: Adapted from Philpot and Nickerson, 1991.

Box 7.3. Environmental or Coliform Mastitis.

Coliform mastitis caused by *Escherichia coli* (E-coli), *Klebsiella*, *Entrobacter* and *Citrobacter* species occurring largely during the first two months of lactation results from the teats coming into contact with an unsanitary environment between milkings. The coliforms do not move from cow to cow but from the cows' environment. When infected, there is a rapid onset and symptoms which include fever, off feed, diarrhea, and shivering. An infected quarter becomes hot, swollen and sensitive to touch. The milk becomes an abnormal secretion.

Frequently coliform infection causes a loss of function in the affected quarter(s). Recovery, minus a quarter generally occurs, however, about 25% will die or need to be culled because of the infection.

Support therapy using anti-inflammatory drugs and injection of oxytocin followed by milking out the infected quarter as soon as possible is recommended.

The New York State Mastitis Control Program recommends:

- Keep stalls clean
- Clean and dry teats before attaching the milk unit
- Provide clean environment for cows at calving
- Clean teat ends and swab with alcohol before any udder injection
- Check milking system regularly to ensure it is functioning properly

Source: Hoards Dairyman, 1993.

---

will be 15 to 40 subclinical cases. Most clinical cases are preceded by infections at the subclinical level (Philpot, 1991). Subclinical cases do not permit visual detection and can not be determined by udder massage unless the infection is advancing to the clinical stage. It is the subclinical cases that are the "hidden losses," because milk yield is lowered. Additionally there are the adverse effects on milk composition and processing qualities (Box 7.5).

Box 7.4. Forms of Mastitis.

- **Clinical** - visible abnormalities in the udder and milk

- **Acute** - sudden onset of redness, swelling, grossly abnormal milk and reduced yield (often caused by coliform)

- **Peracute** - includes above symptoms, but may also include depression, raised pulse and respiration rates, cold extremities, dehydration and diarrhea

- **Subclinical** - not detected visually; must detect presence of infecting microorganisms or products of inflammation, such as somatic cell count (15 to 40 times more prevalent than clinical form, reduces milk yield)

- **Chronic** - may begin as any of the clinical forms; evidence by intermittent signs of clinical mastitis

- **Nonspecific or aseptic form** - occurs when microorganisms cannot be isolated from milk samples, may be clinical or subclinical

Source: Adapted from Philpot and Nickerson, 1991.

The conclusion is that to keep mastitis at a low level it is necessary to direct efforts to keeping the rate of new infections as low as possible and shorten the duration of those infections which do occur.

Box 7.5. Effects of Subclinical Mastitis on Milk Quality.	
• Lactose	Decreased 5 - 20%
• Total proteins	Decreased some
• Casein	Decreased 6 - 18%
• Immunoglobulins (bad)	Increased
• Solids not fat	Decreased up to 8%
• Total Solids	Decreased 3 - 12%
• Fat	Decreased 5 - 20%
• Lipase (bad)	Increased rancidity
• Sodium (bad)	Increased
• Chloride (bad)	Increased
• Calcium	Decreased
• Phosphorus	Decreased
• Potassium	Decreased
• Trace minerals (bad)	Slight increase
• Cheese	Decreased Curd strength
• Heat Stability	Reduced

Source: Adapted from Philpot and Nickerson, 1991.

Methods of Detection

Methods of detection are classified as those that can be executed during the milking process or in the laboratory tests (Box 7.6).

Once microorganisms enter the teat canal, the first line of defense is encountered. All milk contains somatic cells, including primarily leukocytes (white blood cells) and some secretory cells. Leukocytes are produced to combat most health disorders and are extremely important in combating mastitis. When one or more quarters of an udder become infected, large numbers of leukocytes migrate to the infected quarter to destroy and remove bacteria or toxins they produce. This means that a high somatic cell count (SCC), greater than 400,000 in the milk, gives a strong indication of the presence of infection. Factors, such as late stage of lactation, old age and climatic stress can elevate SCC but only slightly compared to the infection from mastitis.

Box 7.6. Detecting Presence of Mastitis.

<u>Cowside Test</u>

* Physical examination - check udder before and after milking
* Strip test - first streams during preparation of udder for milking (strip cup)
* Paddle tests - California Mastitis Test (CMT) - use to estimate somatic cell count just before milking
* Somatic cell Counts - automated electronic cell counters conducted in central laboratory on milk samplings
* Electrical conductivity - One of the new measuring devices that aids in detecting mastitis and to determine susceptibility to antibiotics

<u>Laboratory</u>

* Culture of milk samples - valuable for types of microorganisms causing infections
* Herd level of infection - can use herd milk for somatic cell count, total bacteria and types of bacteria
* Prostaph antibody test - can be used to defect presence of *Staphylococcus aureus* antibodies in composite samples

Source: Adapted from Philpot and Nickerson, 1991.

A feature of the Dairy Herd Improvement (DHI) program in the U.S. is that the milk samples taken for determining milk composition by supervisors during their monthly visits to member farms are processed at central laboratories for individual cows (composite of all quarters) and the results returned to the farms. The well established relation between high SCC and lowered milk yield has led to wide adoption of the cell count feature. Dairymen in Puerto Rico and Taiwan have access to similar programs. Since somatic cell counts are in large numbers, counts by intervals are converted to a linear score (Box 7.7) for convenience of reporting.

Even though no specific SCC or SCCS can be used to differentiate infected from noninfected cows, about 50% of the cows with a SCC >283,000, or SCCS 5 or greater, (Box 7.7) have infection in one or more quarters. As SCCS rises, the proportion of cows and quarters infected also rises. The primary value of routine SCCS for a herd is monitoring a mastitis control program, hence SCCSs should be just that and not used as the only basis of treatment with antibiotics. Research has shown that treatment of most subclinical infections during lactation can not be economically justified as the cost of treatment and discarding milk outweighs the benefits.

SCC for heifers in the first lactation generally range from 20,000 to 100,000 (SCCS 0-3). The SCC are generally higher during the first 2 weeks postpartum due to the presence of colostrum and stress associated with the onset of lactation. SCC of milk from infected cows will increase throughout lactation while the SCC of milk from non-infected cows should not rise.

Monthly determinations of SCC on individual cows can be quite useful. New products are entering the market that can also be used for screening infection level. Milking machine manufacturers are adding electronic measuring during milking. There is a system called 4-WARN which claims that using a calorimetric monitoring for change in milk about 5 days after calving can be used to determine the presence of stress. The level of stress serves as an indicator in forecasting mastitis or other infections. Another is the WESCOR Mas-D-Tec portable mastitis detector made by WESCAR at Logan, Utah. It measures the electrical conductivity of small milk samples. The claim is that the procedure requires only 20 seconds. It is low in cost, $185.00. These are but a few examples, but the main point is that the demand on the part of dairy operators for monitoring infections is rising rapidly everywhere. This in turn has markedly increased research on quick and accurate methods.

Box 7.7. Linear Scores (SCCS) by SCC Intervals.	
<u>SCCS</u>	<u>Range of SCC</u>
0	0 - 17,000
1	18,000 - 34,000
2	35,000 - 70,000
3	71,000 - 140,000
4	141,000 - 282,000
5	283,000 - 565,000
6	566,000 - 1,130,000
7	1,131,000 - 2,262,000
8	2,263,000 - 4,525,000
9	> 4,525,000

Source: Adapted from Philpot and Nickerson, 1991.

Bulk milk analysis for individual herds carried out in laboratories can be used for estimates of infection in herds (Box 7.6). A major value is that culturing of bacteria can serve as indicators of total bacteria counts in the herd and general level of sanitation in the environment. Laboratory total bacteria counts are used by milk processors and health officials to penalize or reject high bacteria milk, e.g., >750,000 in the U.S. The laboratory culturing is also highly useful in identification of types of bacteria which can be very important in treatment of cows, e.g., whether staph or coliform.

Treatment

Treatment of all quarters of all cows following the final milking of lactation with a commercially available dry cow treatment product is widely recommended. Among the advantages are: treatment reaches all infected quarters; serves as effective means in preventing dry period infections; cure rate is about twice as high as treatment during lactation; clinical mastitis at parturition is reduced; and salable milk is not contaminated with drug residues.

No more than 0.5% of a herd should be under treatment for reasons of clinical mastitis at any given time. In well managed herds, clinical cases should be on average no more than 1% of the milking herd per month (Philpot and Nickerson, 1991).

The Mastitis Panel of the American Veterinary Medical Association recommends the following procedures in treating clinical cases:

- treat for 1 day after a favorable response has been obtained; always treat for a minimum of 3 days
- protect cows from extreme weather conditions
- alternate the application of hot and cold water to the udder accompanied with massage and stripping
- use oxytocin at 1 to 8 hour intervals accompanied by stripping of infected quarters

Failures in treatment against clinical mastitis are common and may arise from delay in treating, poor selection of drugs and dose levels, stopping treatment too soon, resistance of organisms and presence of scar tissue in the udder which protects the mastitis organisms from the drug. Cows not responding favorably to treatment and showing repeat symptoms ought to be culled from the herd. Failure by some operators of expensive imported cows to cull chronically infected cows has led to loss of herds.

To estimate the chance a cow will have a second clinical case during lactation, divide the number of cows with three cases by the number with two cases. On average 64% of cows with two cases will repeat, thus the prospects for cure are low. When the infection is in one quarter, it is recommended the quarter be dried off and the cow designated for culling as early as practical.

Preparatory to treatment, it is imperative that the teat ends be disinfected by swabbing the teat end for a few seconds with cotton soaked in 70% alcohol or other suitable disinfectant. This avoids microorganisms present on the teat apex at the time of treatment being pushed into the udder. Udder infusions should be made only with commercial products in a single dose container unless administered by a veterinarian.

Acute toxic mastitis is most frequently caused by coliform bacteria. Successful therapy should be directed primarily against the effects of endotoxins that can result in severe depression, progressive dehydration and inability to rise (Box 7.3). Procedures for treatment proposed by Dr. V. L. Carson at the University of Minnesota are:

- milk out the affected quarter every 2 to 3 hours; oxytocin injection can be used to facilitate milk evacuation
- administer large volumes of balanced electrolyte solutions intravenously (oral fluids are not absorbed); 20 liters in first 1 to 2 hours and up to 60 liters over 12 hours; when cow can not stand administer 150 to 250 grams of sodium bicarbonate with the first 3 to 5 liters of electrolyte
- add 500 ml of 50% glucose to first dose of electrolyte
- counteract inflammatory prostaglandins by administering 1.1 mg/kg of Flunxin meglumine intravenously
- administer corticosteroids in peracute toxic cases

Dietary Supplements

Research has shown that dietary supplement of vitamins A and E and selenium can be important in increasing the natural defense of cows through the function or phagocytic competence of white blood cells, transport of antibodies from the bloodstream into the udder and on the incidence of new udder infections. Vitamin E supplement and selenium for 60 days prior to first calving reduced infections with staphylococci and coliforms decreased 42%; clinical mastitis was down 57% during early lactation and 32% throughout lactation.

Considering the monetary returns shown to emerge as a result of following a comprehensive of mastitis control, it is evident that dairy operators gain returns of 200 to 300% per year on investment solely from increased milk production, e.g., in New York State, investment of $20 per cow per year for the control program yielded a net profit of $137. This means that mastitis control does not cost, it pays.

The estimates of percentages for Holsteins in the tropics and the U.S. by SCC or SCCS (show that the proportion of the cows with SCCS ≥ 5 in the tropics is nearly 4 times greater than in the U.S. (Table 7.1). SCCSs greater than 4 are regarded as having subclinical mastitis and high risks of clinical cases. A study in Taiwan of herd levels of SCC showed that from July to September the herds averaged about 10,000 SCC higher per cow than during the other months. This higher level of SCC corresponded to a near 30% rise in this period for clinical cases of mastitis. Clearly more efforts are needed in the control of mastitis in the WCZ. To execute those features of the comprehensive plan to control mastitis requiring emphasis in sanitation practices will cost little if any more, than in the U.S. and other countries (Box 7.2). Disinfectants for teat dipping will cost more in the WCZ because of need to import; nevertheless, when all factors are considered, such as loss in milk sales, lowered feed efficiency and culling of cows, better mastitis control will pay.

Table 7.1. Percentage of Holsteins in the tropics and the US by somatic cell count (SCC).

SCCS	SCC (10^3/ml)	Tropics (%)	U.S. (%)
0	0 - 18	11	17
1	18 - 35	16	22
2	36 - 75	11	20
3	74 - 141	12	19
4	142 - 283	13	12
Subtotal		63	90
5	284 - 565	11	6
6	566 - 1130	10	3
7	1131 - 2262	6	1
8	2263 - 4523	5	< 1
9	4524 - 9999	5	< 1
Subtotal		37	10

Source: McDowell, 1991b.

Some points for ponder as herd operators recognizing mastitis
is a problem are enumerated in Box 7.8.

Box 7.8. Points for Ponder in Mastitis Control.

- For reduction of mastitis resulting from
  environmental microorganisms, surroundings of the
  cows should be kept as clean and dry as possible.
- Materials in comfort stalls are serious sources of
  infection; dry inorganic materials, e.g., washed
  sand or crushed limestone are preferred.
- Accumulation of manure in exercise lots is a major
  source of infection; near daily pick up of manure is
  recommended and periodic cleaning with a new dirt
  surface is a must.
- Dry cow therapy is effective in combating
  environmental streptococcal infections but is not
  effective against coliforms.
- Wet milking must be avoided, udder and teats
  should be dry.
- Use predipping to prevent new infections.

Source: Adapted from Philpot and Nickerson, 1991.

CHAPTER VIII

MANAGING THE DAIRY ENTERPRISE

Background

Management has become a broadly used term. The context for this chapter will be "Decision making on Husbandry Practices" in which the herd owner or manager must develop a balance between inputs and outputs to yield the best cost-benefit ratios. Skills in management are important because differences in profitability among farms is more related to skills in decision making than from the level of technology employed. All farms in an area may have equal access to needs for good animal health but differences between herds in the frequency of clinical mastitis, high somatic cell counts in the milk, mortality of young stock, rate of involuntary culling and even foot problems will vary by 50%. Reproduction rate is another measure with a wide range of variability among herds in the WCZ, e.g., calving interval 401 to 485 days, age at first calving 28 to 37 months.

In this discussion, certain cautions will be cited relative to recommendations made by a specialist who makes a short visit to a herd and views of herd owners on use of the technology. As an example, the specialist sees breeding efficiency in a herd as a problem and recommends that more time be allocated to observing for estrus. The herd owner nods approval but usually does not execute because he fails to see an acceptable payout for the additional labor required.

These days, a key term is "appropriate technology" for development in countries of the WCZ. Government research organizations react that all phases of technology require verification under local conditions. This has and will continue to be a poor approach because the existing research personnel generally have limited experience in principles of husbandry for commercial dairying, and seldom have any experience in the fundamentals of dairying using improved breeds. A basic question is, where can the private investor who is new to dairying going to turn for guidance on the tools needed to make the appropriate decisions? The counter is that a large pool of technology exists which with a "common sense approach" can be evaluated effectively. In Boxes 8.1, 8.2, 8.3, and 8.4 are enumerated technologies which can serve as needed bases. For instance, in Box 8.1 attention is drawn to the tables on nutritive requirements for cattle during all stages if life. The databases come from the temperate areas, such as the Nutrient Require-ments of Dairy Cattle, prepared by the Committee on Animal Nutrition and published by the National Research Council of the U.S. Although the databases are from outside the WCZ, the author has used these tables on dairy cattle throughout the WCZ and found them quite useful for planning feed budgets and preparation of rations. There are also tables developed on the nutritive value of forages, by-products and grains that can be useful in compounding nutritionally balanced rations.

The listed technologies on improvement of genotype and health (Box 8.2) show that all herd owners are well advised to purchase several publications on animal breeding and animal health to become familiar with the basic principles. The management techniques in Box 8.3 will be largely covered in this chapter but are listed as reminders of factors requiring decision making at the herd level. Points dealing with feed preservation (Box 8.4) are a summation, of the basic principles of forage preservation. All the boxes show that through contacts within and outside countries resource information can be gathered with relative ease and at marginal cost. The bibliography at the end of this book gives numerous examples of publications.

The major objectives of this chapter are to show that: 1) development in skills in decision making is a critical issue; 2) there is a need for animal science technicians to consider cost-benefits in preparing recommendations on applications of tech-nology; and 3) there is a vast array of technology available which can be utilized quite effectively to make dairying with improved breeds economically feasible in most areas of the WCZ.

Herd Replacements

Generally, managers of high producing herds view heifers as the "future of the enterprise" and execute practices to ensure adequate rate of development for calving at 24 to 26 months of age and by this age having achieved 80% of expected mature body weight. In low yielding herds, heifers are treated poorly as they are considered as "no return units," resulting in less than desirable cows. A heifer's eventual contribution to the herd is established at an early age, usually < 6 months.

Box 8.1. Technology Available to Improve Nutrition.

- Tables of nutritive requirements of animals, calves, heifers, pregnant females and lactating cows.
- Tables of mineral and vitamin needs for cattle.
- Tables of nutritive value of grains, by-products, numerous forages and variability in pastures.
- Value of feed additives, minerals, etc.
- Estimates of gain in animal performance through improved feeding.
- Value of heating, grinding, pelleting and alkali treatment of forages.
- Use of sources of non-protein nitrogen with forages and concentrates.
- Means to improve palatability of feeds.
- Preparation of mixes for balancing rations.
- Best feeding strategy for various animal species.
- Laboratory techniques for composition of feeds and forages.
- Value of neutral detergent fiber (NDF) in efficiency of feed utilization and biological efficiency of animals.
- Value of the use of the rumen by-pass protein in ruminant feeding.
- Means of detection of feed alteration or inappropriate preparation, such as effect of silica and meillard reaction on digestibility.
- Identification of phenolic compounds which may inhibit use of plants by animals.
- Limiting characteristics of C_4 grasses.
- Problems of deterioration of feeds under warm, humid conditions.

Box 8.2. Technology for Improving Genotype and animal Health.

1. <u>Genetic</u>

   - Basic principles of genetics and potential for genetic change.
   - Value of systems of breeding (inbreeding, outcrossing and crossbreeding).
   - Breeding plans with flexibility.
   - Heritabilities for supportive traits, including disease resistance.
   - Potential value of embryo transfer.
   - Importance of genotype - environmental interactions in animal performance.
   - Value of sire selection on predicted transmitting ability, pedigree and progeny test.
   - Variance components in milk yield in various environments.
   - Probability of success with crossbreeding.
   - Problems in development of new breeds or strains.

2. <u>Health</u>

   - Value of vaccination program.
   - Endo- and Ecto- parasite control procedures.
   - Value of quarantine, slaughter, sales to control health problems.
   - Importance of sanitation in herd performance and disease control.
   - Recommended herd health programs.
   - Have available a health manual, e.g., The Merck Veterinary Manual to use in checking symptoms observed to identify disease.

Dairy breed calves attain a reasonably stable body temperature within a few hours after birth, hence they rapidly develop a wide tolerance to ambient temperature. Elevated humidity levels in calf quarters, due to lack of ventilation or frequent wetting of floor surface, is much more important than temperature for adversely affecting health. Fortunately, the use of "outdoor calf hutches" for individual calves from 0 to 3 months is replacing the "closed barn" or calf quarter thereby reducing both morbidity and mortality from lung damage due to high humidity levels and

---

Box 8.3. Management Techniques.

- Factors leading to mastitis.
- Value of udder and teat sanitation.
- Recommendations on use of teat dip.
- Effect of nutritive imbalances on biological efficiency of cattle.
- Pastures: weed control, use of fertilizer, stocking rate, relation of plant maturity to nutritive value, rotational grazing, nutritive value by season.
- When to offer supplement on grazing.
- Estrous detection to improve breeding efficiency.
- Results of laboratory analysis of feedstuffs.
- Need for care of female calves.
- Appropriate rate of development for heifers.
- Minimum growth rate to ensure "normal" animals at maturity.
- Value of attention at time of parturition.
- Value of foot care.
- When to breed cows and heifers.
- When to assist in parturition.
- Diseases affecting reproduction.
- Drying off requires care.
- Effects of low rumen pH on production and health.
- Effects of aflatoxins in feeds.
- Value of routine somatic cell counts in herd management.
- Value of progesterone in milk samples for management of breeding.
- Publications on guidelines for calf rearing, feeding during pregnancy and care of the dry cow.

reduction in mortality. In the US outdoor hutches can be cheaper than barns, about $100 for a plywood hutch up to $300 per commercial hutch, plus site preparation. An enclosed facility with adequate ventilation will cost $600 to $1,000 per stall.

1. <u>Growth Rate</u>

Recommended practices for feeding calves are projected to more than double birth weight (>225%) by 3 months of age. A delay in doubling birth weight until 6 months will not generally deter attainment of near normal size later under good feeding but if birth weight is not doubled before 8 months, there will be permanent effects on mature body size both in weight and skeleton dimensions but most significant will be lowered response to feeding during first lactation (McDowell, 1985). Level of feeding and care to ensure thriftiness in calves is more important than maximizing rate of growth, e.g., heifers should not build up fat deposits, particularly before 12 months of age. The recommended management system is to have dairy breed heifers attain 50% of their expected mature weight by 15 to 17 months and to reach 75 to 80% of mature size at the time of first parturition. With birth weight ordinarily 6 to 7% of mature weight, a modest rate of gain, approximately 0.5 kg per day, is required for Holsteins from birth to reach 80% of mature weight and about 95% of maturity in skeletal dimensions by 28 months (McDowell, 1985).

At North latitude 38° or higher in the U.S., Canada and Western Europe, birth weight of Holstein females is 40 to 44 kg (Table 4.1). They average 350 to 375 kg at 15 to 17 months, the expected time of breeding, 490 to 525 kg at 24 to 26 months and 625 to 680 kg at maturity. In latitudes less than 34° (subtropics) in the U.S. and Latin America, Holstein females weigh 6 to 10% less at birth (36 to 38 kg) and near 20% less at maturity (Table 4.1). However, in the tropics birth weights may be only 20 to 25 kg when the dams are poorly fed during gestation. Such should be avoided as Holsteins 35 to 40% below normal in weight at birth will likely not reach expected weight at maturity.

2. Management Systems

As recommended in Chapter III, a feed budget for developing appropriate herd replacements is mandated. To optimize rate of body growth, several factors should be considered, e.g., calving weight, breeding age and calving age. In general, chronological age and physiological age of heifers are perceived as analogous to each other, but such is biologically invalid as physiological age depends on mature weight, therefore is not fully dependent on time. Cows should be seen for example as 550 kg animals in most of the WCZ achieving a certain percentage of this expected weight by a given period of time.

The next stage is to determine within available resources how best to achieve the desired growth rate for first breeding and calving - rapid, intermediate, or slow. Rapid development (225% of birth weight by 90 days) requires good quality feeds. Cattle have a capability for compensatory growth which can be used when required (Figure 8.1). Compensatory growth is defined as a phenomenon manifested by animals which after a period of nutritional deprivation, which restricts growth, are provided sufficient quantity of feed energy or other deficient nutrients to grow at a more rapid rate than do unrestricted animals of the same age and size. Animals suppressed by low energy intake will undergo partial dehydration. Upon realimentation, appetite

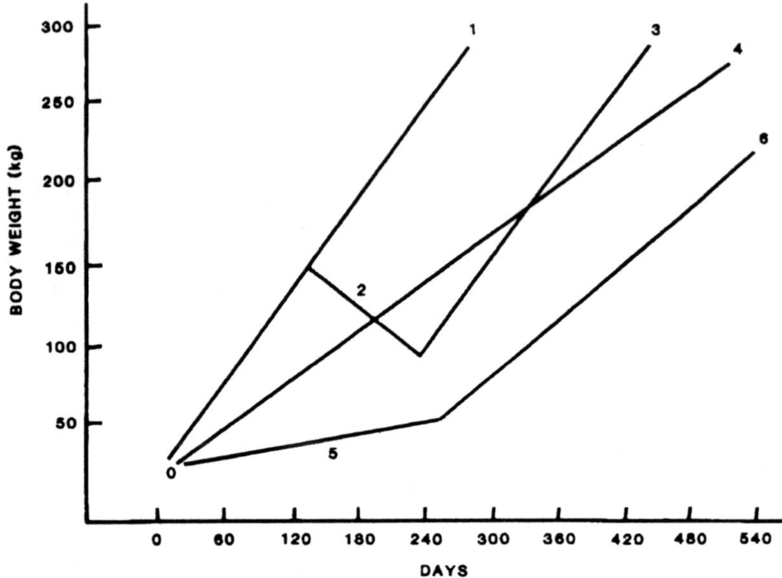

Figure 8.1. Growth rate of Holstein calves from birth to 540 days under various management systems; line 0-1, high feeding, rapid growth; line 2, rapid growth to 100 days followed by depressed feeding; line 3, illustrates compensatory growth following period of weight loss; line 4, medium growth ate (recommended WCZ); line 5, very slow growth; line 6, slow growth with low response to good feeding; line 5-6 will result in abnormal heifers (Adopted from McDowell, 1985).

increases, water content of body tissues rises and so does protein deposition. The higher level of appetite lasts for 30 to 45 days or until the animals have reached normal metabolic levels. A period of compensatory growth in cattle usually follows regrowth of pastures after the wet season starts. It can be employed in heifer rearing but requires careful management to achieve the desired end results.

In Figure 8.1, the line designated a "1" shows the capability of Holstein females to respond to high feeding; not highly recommended as can lead to early deposits of fat leading to risks of decreasing mammary duct development. Line 2 reflects the common condition when rapid growth calves are placed on pasture at about 180 days of age; loss of weight. Such will not be detrimental provided the heifers receive pasture plus supplement by 200 days so they can make compensatory growth (line 3). Line 4 is the recommended rate of development in the WCZ. The rate of growth is modest but the heifers remain thrifty (condition score of 3.0) and do not accumulate excessive fat. A feeding program depicted by lines 0 - 5 - 6 is undesirable for dairy breeds as heifers will be later in reaching puberty, slow breeders, have high dystocia at first parturition, and will yield milk at least 30% below expectancy. Additionally slow growth is a stress on calves leading to high risks of either diarrhea or respiratory infections resulting in rough or shaggy hair coats which in turn increases the level of stress. In the WCZ, a cow needs full lung capacity to combat heat stress. When permanent lung congestion occurs from an early respiratory disorder, such as pneumonia, performance will be low as a cow. Internal parasite infestation will cause a similar condition.

To minimize risks of illnesses, the neonatal calf should receive colostrum and be kept as dry as possible with measures taken to control flies. Fly control is seldom deemed important but usually is cost effective. In many herds there is an over use of water in the calf quarters leading to high levels of humidity. When the relative humidity exceeds 80%, calves become chilled during the night which increases susceptibility to infections. High fly populations also increase disease transmission.

Heifers need routine treatment for control of internal parasites. Heavy infestations will cause unthriftness and delayed puberty or they may never reach sexual maturity. The author sees too many heifers over two years of age in herds of the WCZ with infantile ovaries from chronic parasites and the resulting diarrhea. If the parasitic heifer does eventually calve, the odds of her being culled for sterility or low production is well above average. Holsteins should reach breeding age at 50% of mature weight by not later than 24 months. When delayed to 28 or 30 months, development of physiological processes is impaired to the point the heifer will show poor response to improved feeding during lactation.

If heifers weighing 265 to 280 kg conceive, they need to increase body weight to at least 440 kg to avoid serious difficulty at calving. The required gain of .60 to .66 kg daily is hardly achievable from grazing alone. Delaying calving beyond 38 months, Holsteins will be low or nil in profitability. The odds for culling for poor breeding efficiency or a health problem before completing two lactations is near 70% (McDowell, 1991b).

Heifers need exercise, especially post weaning to ensure good, sturdy development. After 6 months of age, heifers should be allowed full movement in a pasture or lot for two or more hours per day.

3.   <u>Feeding of Young Calves</u>

For dairy breed heifers the recommended ADG throughout the growth period should be .5 to .6 kg. To start the process, the newborn female should receive 4 kg of colostrum for at least 3 days followed by milk feeding at the rate of 10% of body weight for at least 28 to 30 days. The feeding of colostrum cannot be over emphasized as the increase in disease and death associated with failure to acquire adequate colostral immunoglobulins is related to significant economic losses, not only in the cost of replacement heifers, but also the expense for treatment and extra labor involved in caring for sick calves; (See also Section in Chapter VI, Problems in Intensive Enterprises). The first milking following calving contains the highest content of immunoglobulins. Therefore, this milking should be saved for feeding of the newborn calf over the first 48 hours of life.

Commercial dairies usually wish to maximize milk sales, thus they shift to a calf starter mix and forages are offered at about 30 days of age. These feeds for calves can be purchased in most locales but are largely based on imported ingredients. Local mixes can be prepared but their nutritive value needs to be known for preparation of satisfactory rations. In Table 8.1 are the nutrient composition of feeds for young calves and growing heifers to serve as guidelines. Both the milk replacer and calf starter must have high densities in energy and protein with low fiber. Adherence to use of satisfactory feeds, the calf becomes ready for weaning to pasture and other sources of coarser feed at 6 months and should have reached 150 kg or more in body weight. A number of tests in temperate areas show milk feeding can be restricted to once daily and weaning from milk at 28 to 30 days of age, then shifting to a milk replacer or to a commercial calf starter containing 18% CP will provide acceptable rates of development to about 60 days. A fiber-limiting complete feed formulated with 12.5% CP fed free choice along with some hay will be satisfactory up to 180 days. Another caution on young calves receiving high liquid diets (milk or milk replacer) is to prevent the sucking habit. When calves are being reared in groups, sucking can become a real problem; ears, tails, navels or other objects. When calves are fastened up for feeding, duration should be at least 15 minutes post feeding and the separation great enough to prevent reaching each other's ears. One of the good features of using individual outdoor hutches is that problems of sucking are markedly reduced.

As pointed out in Chapter III, grazing alone or green cut tropical forages will not generally be adequate to support a sustained gain of at least .45 kg per day. For supplementation of forages, the suggested concentrate mix for lactating cows (Table 3.10), containing 64.3% TDN, 18.2% CP, > .57% Ca and .45% P has served as a suitable supplement after 6 months of age. The recommended feeding rate is 2.5 kg per day to 18 months and then the level raised. When green chopped grasses like Napier are the forages, heifers should be offered at least twice (preferably three times) their level of intake over 4 to 6 hours per day in order to permit selection or otherwise the concentrate supplement will need increasing by 25%. Additionally, hay or chopped straw should be provided at the rate of .5 kg per day from 6 to 12 months and 1.0 kg daily > 12 months to ensure establishing and maintaining satisfactory rumen pH (See Section Green Chop, Chapter III). The supplementary concentrate should be fed separately, not poured over the forage, as a means of trying to "trick" the heifers into higher consumption. The stems of Napier grass do not contain sufficient TDN to support the needed .45 ADG. However, the level of supplementary feed should be such that heifers will not substitute for the forages which they are quick to do as they do not have the drive to eat forages like lactating cows.

Table 8.1. Nutrient compositions for calves and growing stock.

Nutrient	Calf milk replacer	Calf starter mix	Growing heifers		
			3 to 6 mo	6 to 12 mo	>12 mo
Energy					
ME, M cal/kg	3.78	3.11	2.60	2.47	2.27
TDN as %DM	95	80	69	66	61
Protein, CP%	22	18	16	14	12
Fiber, NDF%	-	-	23	25	25
Minerals					
Ca%	.70	.60	.52	.41	.29
P%	.60	.40	.31	.30	.23
Vitamins					
A, IU/kg	3800	2200	2200	2200	2200
D, IU/kg	1200	600	300	300	300
E, IU/kg	40	25	25	25	15

Source: Adapted from NRC, 1988.

Risk of substitution of supplements is another reason the forage needs to be of good quality. Experiences in Puerto Rico using fertilized pastures of improved grasses were that 30 kg of supplement was used per kg of gain per heifer over grazing alone which is not economically viable. When the heifers can select grass with 55% digestibility (TDN) with a CP content of at least 10% supplement will not pay. The best way to monitor the need for supplement on pastures is for managers to continuously assess body condition score (Figure 5.1, 5.2). When the heifers are below a score of 3, supplement is recommended.

These guidelines are oriented to obtain heifer development to at least 450 kg (preferably 475 to 480 kg) and calve at 28 to 30 months with few difficulties.

4. Dehorning and Removal of Extra Teats

For herds kept either in partial or full confinement, dehorning is strongly recommended. Removal of horns at 1 to 3 weeks of age is considered best. When done later, there is bleeding causing high risk of infection in the frontal sinuses.

Successful dehorning can be achieved in young calves using a commercially prepared stick of caustic paste, or with an electric iron, cube or Barnes dehorner (Hoard's Dairyman, 1990). When using the caustic paste, the hair should be clipped 2.5 cm around the horn base followed by applying the caustic stick on the clipped area. The clipping is mainly to determine location and size of horn button. Some vaseline, lard or other fat should be applied in the clipped area to avoid the caustic running over the face and causing burning. The end of the moistened caustic stick needs to be rotated around the button until all parts are treated equally. The skin is soft so small bleeding will become evident quickly; the point to stop.

There are a number of makes of electric dehorners which operate on the same principle. The secrets of the electric is proper heating and application long enough to destroy the active horn tissue, about 10 seconds when heat is right (Hoard's Dairyman, 1990). No matter the method used, the head of the calf should be tied up for approximately two hours to avoid contacts. If it is not possible to dehorn calves earlier as described, tube or Barnes dehorners can be for used when the calves are about 4 months of age.

Young calves of dairy breeds should also be checked for extra teats. It is recommended these be removed. First wash and disinfect the teat area, then stretch the teat and snip off close to the udder followed by application of more antiseptic solution. A caution is that the "extra teat" is the one removed.

5. <u>Imported Heifers</u>

Comments are offered on imports because of the need to protect the high investment and for durability to produce in the future.

The usual preference of purchasers is to obtain heifers guaranteed as sound breeders (pregnant) in order to obtain quickest returns on their investment; yet, the transfer of Holsteins from their initial temperate environment at 4 to 6 months into pregnancy places them under tremendous stress. Such often leads to abortions, weak calves, numerous pre-and postpartum health disorders and slow rebreeding during first lactation. Over the short range, the importation of pregnant heifers gives highest returns, but for the long range (> 5 years) importing younger heifers (12 to 14 months) will likely yield highest returns. The younger heifer becomes adjusted to the local environment, the feeds, management regimen and develops resistance to a number of local health problems before late pregnancy. These may include organisms causing foot rot, various internal parasites, and diseases that would otherwise affect later performance. With an opportunity to adjust, the younger imports will have a more productive life. This, of course, delays initiating receipt of income, therefore a trade-off on short versus long-term gains.

The major oversight by importers, be they private investors or government institutions, is not "preparing the environment" for the imports. The greatest shortfalls are: lack of knowledge of quality of local feeds particularly forages, and the needs in nutrients for acceptable biological functioning, i.e. to reduce stress; low skills on the part of local labor to handle the high level of technology in the genetics of the heifer; and inept sanitation practices for manure disposal, fly control measures, and wetness both inside and outside buildings. In both Puerto Rico and Mexico, it took about 10 years for buyers to learn how to handle imported pregnant heifers to reduce morbidity losses, such as being separated from the rest of the herd for the first 60 to 90 days, care in feeding and adjustment by the heifers to the tones of the local language. The latter would not be assumed as a problem but language can be a strange element which causes excitement and secretion of adrenalin.

Because of the stress of transport and surroundings, the incidence of dystocia can be above average. Such suggests careful watching during parturition. Should assistance be needed, then exercise care as described in Chapter VI on calving difficulty. The incidence of udder edema is normally above average in first lactation but can be even greater in imports. Why, we are not certain but it is assumed to be another product from the changes in environment.

The bottom line is that imported heifers represent a large investment which should be meticulously observed at all angles and actions taken accordingly.

The Lactating Herd

Chapter III on Feed Budget dealt with the "technical aspects" on the limits of forages and needs for supplementary feeding to achieve an acceptable biological efficiency of dairy breeds. As an example, values in Table 3.5 show that for a Holstein cow weighing 550 kg and expected to produce 6,000 kg of milk during a 10 month lactation, 78% of the total TDN required will need to come from sources other than green chopped King grass. This declines to 60% during month 10. The intent of this section is to put forth suggestions on relating the use of feeds to the cows.

1. <u>Feeding</u>

In Box 8.5 are proposals on feeding lactating cows which have proven beneficial to managers seeing the best cow responses.

Box 8.5. How Much Should Lactating Cows Eat?

- Should reach maximum DM intake no later than 10 to 12 weeks postpartum.
- Should eat at least 3% of body weight per day, e.g., 525 kg cow x 3% = 16 kg DM.
- For each 1 kg of expected milk, should eat 0.5 kg DM; eating more causes excessive body condition and metabolic disorders.
- Cows want to eat after milking so adapt feeding to eating behavior.
- When possible, provide separate milking/feeding for heifers as they in first lactation eat 15% more when housed separately.
- Best total ration DM is 50 - 75%. When silage is heavily fed, expect DM to be decreased by .02% of body weight for each rise 1% in ration moisture.
- Expect cows to decrease DM intake about 3% per 1.2°C rise in daily ambient temperature >24°C.
- Provide clean water within 20m; expect cows to drink about 2.0 liters for each kg of milk.
- Feed at least 60% of the ration at night.
- Silage pH should be less than 4.2 for maize, less than 5.0 for legumes.
- Feed should be available to cows up to 20 hours per day.
- Provide each cow 0.5 to 0.6 m of feeding space.
- Feed manager height is best when cows are eating with their heads in downward position (grazing level) to minimize wastage and obtain highest intake.

Adapted from: Mahana, 1990.

These are adapted from 60 odd "rules of thumb" that nutritionists and herd managers have found to help ensure lactating cows are offered feed nutrients when returns are expected to be best (Mahana, 1990).

Particular attention is drawn to the last of the listings, feed manger height, since it plays an important role in rumen function. Cows with their heads in a downward position while eating produce 17% more saliva than feeding with the head in a horizontal or raised position. In the WCZ, we need a high rate of saliva secretion to maintain proper rumen function. Also when the feeding is from elevated bunks, cows exhibit more rooting, sorting, and feed tossing behavior leading to feed wastage (Albright, 1993). Other worthy considerations are: social mixing (group feeding) influences eating bouts and maximizes intake. Palatability can have a major influence on feed intake as the sense of taste is well developed in cattle.

2. Peak Yield

It is difficult to assess how the lactating cows are adapting to the herd feeding regime. Two aids in this respect are checking out peak yield (Box 8.6) and body condition score (Box 8.7). In temperate areas cows reach peak milk yield 8 to 10 weeks postpartum. Currently in most of the WCZ pure types or crosses with a high proportion dairy breeding are peaking at 4 to 6 weeks. We would like to extend the peak period, hence a goal. Peak yield is deemed important as generally each extra kg of milk can add 100 kg to lactation yield. Suppression of intake from heat stress and marginal level of protein in the ration appear the main restrictions. Hence expected time of peak yield is a good period to evaluate the daily feed program. The guidelines listed in Box 8.6 should prove helpful.

3. Condition Score

Body condition score is a useful tool for managers to monitor their cows. Assessments are made following the photographs in Figures 5.1 and 5.2. Some guidelines in relating condition score to what is seen in the herd are in Box 8.7. The desired condition score at parturition is 3 to 4 but often dairy breeds score 4 to 5 which is over fattening in herds of the WCZ. Cows scoring 4 or greater before calving lose more subcutaneous fat. In improved breeds these reserves are used for milk production without negative effects on reproductive performance or on the incidence of retained placenta. However, overfattening can lead

Box 8.6. Look at Peak Milk Yield.

- Cows should peak 8 - 10 weeks postpartum; WCZ 4 to 6 weeks.
- First lactation cows should peak within 25% of older cows.
- Each extra kg of milk at peak adds about 100 kg to lactation yield.
- If cows are not peaking as expected, check protein; if peaking but persistency not good, check energy level.
- After peak, heifers will decline in milk 0.2% per day, cows 0.3%.
- Forage DM intake should be 2% of cow's body weight.
- Make certain 22% NDF in total ration.
- Forage NDF should be 0.9% to 1.0% of body weight; with chopped grass, need 2 to 3 kg of dry hay or chopped straw.
- Provide at least 2.2 kg of fiber over 4 cm long; underfeeding of effective fiber causes off-feed and low percent fat in milk.
- Rumen pH should be > 6; lower pH decreases fiber digestion, limits protein synthesis and increases potential for acidosis.

Adapted from Mahanna, 1990.

to a rise in metabolic disorders such as ketosis. The author is finding condition score quite useful in both quick assessments when the operator is unsatisfied as well as in dialoging on proposed improvements in feeding and management. For condition scoring in a herd, it is recommended that periodically someone not directly associated with the herd do an appraisal. This is because those closely involved with a single herd tend to use the average for cows in the herd thereby narrowing the range on scores.

4. <u>Use of bST</u>

Bovine Somatotropin (bST) has been widely tested in temperate areas for use as a stimulus for milk production, increasing lactation yields 15 to 20% (Bauman, 1992). Usually bST is injected or implanted commencing 60 to 80 days postpartum. Although bST is being used in commercial herds in Mexico, it is not recommended for general use in the WCZ because more feed is needed by cows receiving bST to make it effective. Where temperatures suppress appetite and the feeds are rather coarse, the effects would probably be low or could be detrimental in the early part of lactation. Use of bST also has the risk of slowing of rebreeding.

However there is another possible scenario to derive a cost-benefit plus in the use of bST. Testing of practicality has not yet been carried forward to conclusive results, but the rationale is sound. The objectives are to increase milk yield mainly by extending days in milk, reduction of days dry and calving interval, and lowering metabolic disturbances at the following parturition. An example follows:

A. Averages for Herd

* milk yield per lactation 4690 kg
* days in milk (DIM) 280
* days dry (DD) >150

- calving interval (CI) 455 days
- daily milk at 200th day of lactation is 12 kg
- daily milk at the 280th day of lactation is 6 kg

B. Use of bST - start treatment at 200th day to maintain higher yield and to delay the rapid decline which generally occurs after about 200 days. For example:

Day of lactation	Yield/d (kg)	
	Present	Yield with bST
200	12	14.4
230	10	13.2
260	8	12.0
290	6	9.0
320	8	8.0
350	0	7.0
380	0	6.0

With CI 455 days and days dry > 150 days, bST should raise income and avoid over conditioning of the cows. If rebreeding is slow and bST used up to the 380th day of lactation, concentrate needs would rise by 440 kg/cow and milk yield by 918 kg with CI 450 days or 708 kg with CI at 420 days. Following calculations and council with the best researchers on bST, it is projected that the use of bST can have considerable advantage in herds of the WCZ towards improving net return through extension of the time for profitable milk production.

5. Dry Period

The recommended length of dry period is 60 days. Shorter periods do not permit sufficient time for regeneration of mammary tissues; while, long periods leads to over fattening, including fat deposits in the udder, which can reduce later production.

The most accepted method of drying off is to stop milking abruptly. Buildup of intramammary pressure inhibits further milk secretion. Following the last milking, all cows should be treated in all quarters with an effective dry cow mastitis product. Dry cows should be watched the first two weeks after drying off and two weeks prior to expected calving. Cows with quarters that appear abnormal should have the quarter milked out and retreated.

Place dry cows in a separate area with feed and water and space for exercise. Cows remaining with the lactating herd may overeat and develop disorders at the next calving. There should be a dry sanitary place for resting to minimize risk of mastitis infection. In the WCZ, calving outside in an open lot is found to be best.

Dry cows should be fed 1.5 to 1.8% of body weight in DM daily (550 kg Holsteins 8.5 to 10.0 kg). The overall ration should be at least 55% TDN and 12% CP. Feeding can be once daily but when chopped forage is used feed should be at least twice daily.

Foot Problems

We have problems in maintaining "good feet" with improved dairy breeds in the WCZ, particularly when in confinement. Problems may be caused by inheritance but genetic effects are kept generally low as sires with defective progeny are quickly removed from service, therefore the major causes are the result of the environmental conditions from continuous wet floor surfaces, low sanitation, infectious organisms or nutritional imbalances (McDaniel et al., 1984). Few dairy operators in the WCZ as yet recognize that cows are only as good as their feet hence do not exercise care on a systematic basis. A study conducted in New York State found that trimming annually would extend herdlife of cows by at least one year.

Factors often affecting condition of the feet of lactating cows are enumerated in Box 8.8. Visual appraisal by experienced personnel can use the suggested guidelines in Box 8.9 and Figure 8.2 to determine whether there are problems arising. Using these guides, a herd manager may conclude that near half the herd has claws with lower angle than the recommended toe

‖ **Box 8.8. Factors Affecting Cows' Feet.**

Floor rough	- extensive wear when wet
Floor smooth	- too little wear
Wet floor	- soft hoof, infection, sole ulcers
Dirt surfaces less than 2 h/d	- foot wear too rapid
High concentrate to forage in diet	- sole ulcers
Low neutral detergent fiber level	- acidosis, lameness, sole ulcers
Rapid change to high energy diet	- lameness, sole ulcers

‖ **Box 8.9. Assessing Foot Problems.**

1. Appearance (recommended):

Toe angle, degrees	40 to 50
Toe length (mm)	
1st lactation	60 to 70
Later lactations	60 to 80
Diagonal length (mm), top of	
heel to end of toe	140 to 150

2. Stance on rear legs (Figure 8.3):
   A. Straight - equal weight all toes.
   B. Outside claw overgrowth, legs spread apart.
   C. Outside claw more overgrowth, feet spread, knees close together.
   D. Outside claw badly overgrown, feet wide, knees together, difficulty walking.
3. Ulceration of foot sole, cow slow to move, evidence of ulceration when seen from bottom (Figure 8.2).
4. Infection between toes - toes spread, redness in skin coloration above toes, lameness.

angle 50°, greater foot length than 70 mm and diagonal foot length > 150 mm. On average for herds in the WCZ, >70% of the cows will exceed the norms.

When there is over growth causing the toe angle to decline and foot length to increase, more weight pressure is placed on the heels which creates discomfort and difficulty in walking or reluctance to mount other cows in estrus or be active when in estrus. Another serious condition can result from unequal growth of the outside claws of the rear feet as shown in Figure 8.3. The usual in herds of the WCZ is that about 40% of the

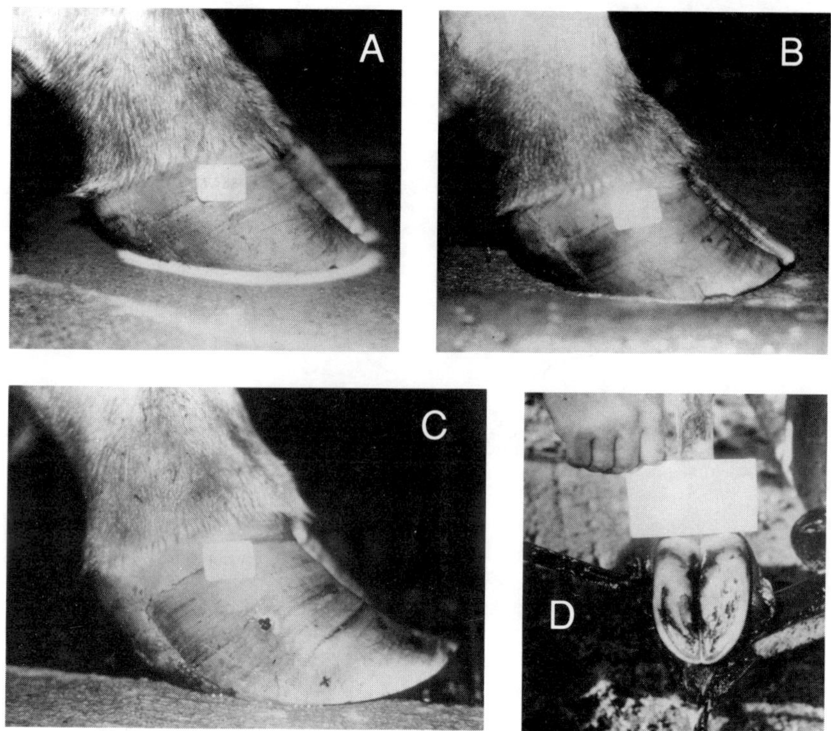

Figure 8.2. A - good hoof (toe angle 40 to 50°, toe length 60 to 70 mm, foot diagonal 140 to 150 mm, top of heel to end of toe, Box 8.9); B - some elongation with excess weight shifting to heel (time to trim); C - low foot angle, toes turning up, cow uncomfortable to walk on heel; D - foot pad elongated and showing ulceration; cow became crippled (Courtesy B. T. McDaniel, NC State University).

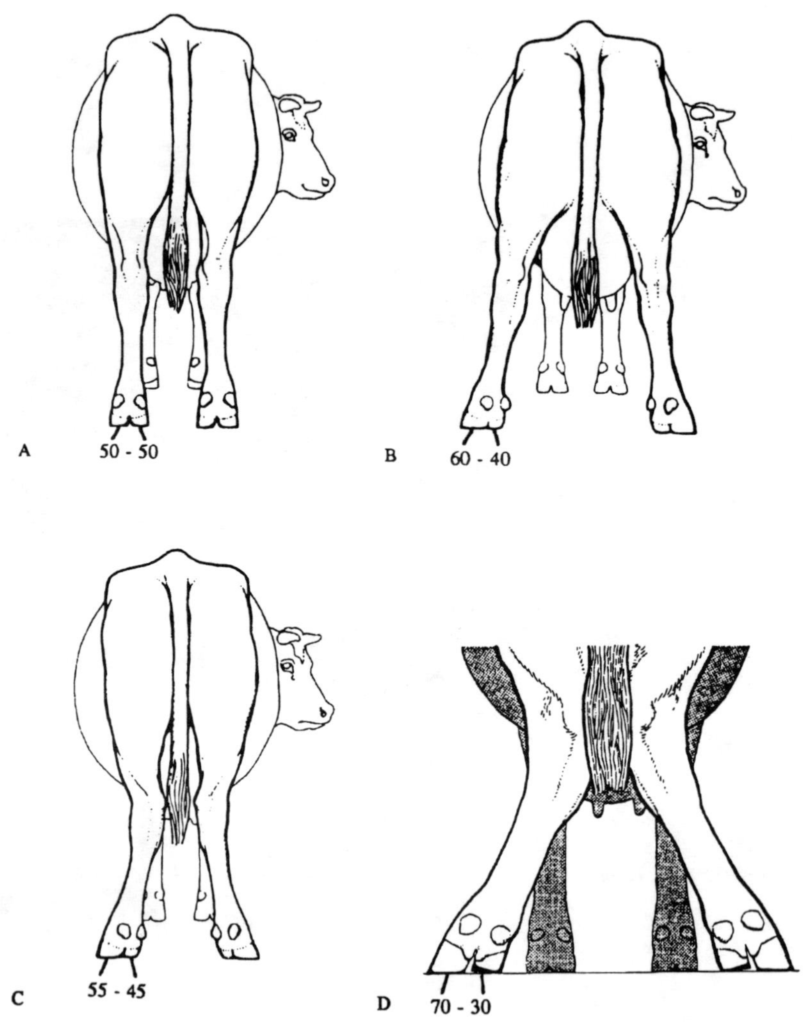

Figure 8.3. Position of stance on rear legs; A, desired, 50:50% of weight on each claw; B, outside claw commencing to overgrow, legs spreading, weight shifting to 60-40%; C, outside claw overgrowth causing hocks to shift inward; and D, deflection in leg angle, cow undergoing serious discomfort because of serious outerclaw overgrowth (Drawings courtesy of Dr. E. Toussaint Raven University of Utrecht, The Netherlands).

cows will score as A, (desired); B, 30%; C, 20% and 10% D. Differences in height develops between the two hind claws results from overgrowth of the outer claw. The stance is influenced by adaptation following overloading problems in too high outer hind claws. The adaptation of the stance only gives a partial unloading. The functional-anatomical phenomenon arises from systematically contusion problems in the cornium of the outer claws with resulting pain causing adaptation of stance. If not treated in time by functional trimming this can lead to development of ulcerative processes in the claw.

Viewing from the side, nearly 25% of the herd will show hunching in the mid-spinal vertebra (raised above the horizontal line) which gives further indications of need for foot care. Additionally, there may be cows showing signs of infection between the toes in one or more feet. If the pads of the rear feet were viewed, no doubt 30% or more would exhibit ulcerations in the soles, such as D in Figure 8.2. When ration imbalance in fiber is poor, causing laminitis, 70% or more of the herd could have ulcerated soles.

The rear feet appear most subject to disorders, but the front feet are also subject to over growth and infections. The observation that frequency of foot disorders can be high during first lactation may be surprising but occurs when heifers are shifted from pasture or dirt surface lots to concrete surface and fed a ration low in fiber.

The over extension of the hooves can be helped by trimming. The investment in tools is not large; the main cost will be labor. The tools are: a 2.5 width wood chisel, a hard rubber or wood mallet, a rasp, a set of hoof nippers and a pair of hoof knives (right and left). Also required is a suitable place for constraining the cow and a one-half meter square piece of plywood for placing the foot when the chisel is used. The same tools and procedures can be employed to reduce extension of the outside toes to renew proper stance. However, there is little that can be done through trimming when cows reach stage "D" in Figure 8.3. Foot trimming will also help to restore the mid-spinal vertebra to normal position.

The next step is to review the factors affecting the feet (Box 8.8) and consider adjustments, e.g., the concrete surfaces in the barn are smooth and the floor is washed down twice daily. The smooth surface slows foot wear and wetness leads to soft hooves along with a tendency for ulceration in the soles (D, Figure 8.3). The dirt exercise lot is muddy and has a heavy accumulation of manure thereby keeping the feet wet at all times. Attention to these aspects of the herd environment will pay dividends, e.g., building of mounds on the exercise lot will help drainage and health of feet.

Table 8.2 provides estimates of foot growth and wear per month during lactation and the dry period for a herd using principally pasture and a herd in complete confinement. Cows from both herds were on pasture during the dry period. Growth was greater than wear throughout but cows on concrete during lactation showed decline in lactation two. The rate of regrowth during the dry period may prevent serious problems while on concrete during lactation. However this shows that use of pasture or dirt lots during the dry period is a desirable practice.

In sum, the author contends that considering cost and required resources, more attention to foot care in the WCZ herds will give high returns, not only from more milk but improvement in rate of rebreeding and duration of herdlife, e.g., 300 kg more milk, 30 day sooner rebreeding and 1.0 to 1.5 more calvings per cow are

Table 8.2. Differences between growth and wear (mm/mo) for herd mainly on pasture and one mainly on concrete.

Hoof	First lactation	Dry period no. 1	Second lactation	Dry period no.2
Herd 1 - Pasture				
Front dorsal	.66	.73	.17	.35
Rear dorsal	1.07	.42	.68	.02
Herd 2 - Confinement				
Front dorsal	.07	1.94	-.91	1.51
Rear dorsal	.15	.37	-.75	1.00

Adapted from Hahn et al., 1986.

easily attainable returns. For example, it has been well demonstrated that cows with steeper foot angles have longer herd life in all housing types.

There are a number of good publications on recommended hoof care, e.g., "Hoof Care for Dairy Cattle" (Nocek, 1993).

Housing

In the WCZ, the norm for commercial herds using dairy breeds is either complete or semi-confinement. Confinement creates a need for different management practices because of the limitations listed in Box 8.10. As yet, there are no recommended models for housing in the WCZ. In general, there is a tendency to invest more capital than necessary and with more interest in "convenience" to labor than comfort of the cows. When importers come to the U.S., they become impressed with automation and wish to duplicate. The equipment manufacturers hasten to impress importers. The net result is that the importers often "over buy." A major mistake is investing in automatic equipment which usually has a short operating life under WCZ conditions of climate (high humidity) and capacity of labor. This section focuses on "speaking for the cows" considering the principles on environment in Chapter II and investment costs.

Box 8.10. Confined Housing Trends.

- Restrict animal's opportunity to seek comfort
- Creates problems of humidity that are more detrimental to health and comfort than high temperature
- Limits opportunity for exercise
- Can increase health problems
- Lowers reproductive efficiency due to problems of heat detection
- Maintenance of sanitation requires higher inputs, e.g., manure disposal
- Problems of social dominance can be greater
- Increases capital investment per cow

From the cow's point of view, she would like to have: some freedom of movement; have access to a dirt surface the major part of the day; surroundings with as high level of sanitation and as dry as possible; feed and water available; and be able to use reradiation of body heat to maintain comfort. Confinement to a covered area with feeding and comfort stalls for resting does not fully support the cows' needs.

There are several misconceptions about using covered shelter at all times that are worthy of consideration in herd management. Examples are: 1) shelters reduce the heat load to some extent during mid-day but in environments where the maximum temperature equals or exceeds 30°C, the number of hours the temperature is 27°C, or higher is the same under a shelter as outside (Figure 8.4), 2) there is high value in use of shadow cast by roofs defined as the partial reduced light within a space from which rays of the sun are cut off, and 3) utilization of sky temperature for radiation of heat and for rapid cooling during twilight and darkness can be highly effective in cow comfort.

A good environment for lactating cows, dry cows or heifers is illustrated in Figure 8.5. A shade or shelter, placed on a raised surface of earth, is located in the center of the lot with free access to all sides. When oriented east to west, the shelter casts a shadow in both the morning and afternoon. Given free choice, most of the cows will seek the shadow cast by the roof on the west side during AM (OOO, location of cows), move underneath during mid-day, move to the shadow on the east side after about 2 PM, and become scattered over the lot during hours of darkness. The temperature of the sky where the sun is located for direct rays striking the cows may be as high as 50°C but the sky away from the sun may be as low as 15 to 20°C which means heat can be radiated to the cool sky if the cows have opportunity to avoid direct rays. Allowing use of the shadow will not reduce the environmental temperature but can be of high value in maintaining comfort. During twilight and darkness, heat gain from the sun is low or nil, hence the cows can restore heat balance in the body much more rapidly outside than under a shelter. Ideally, there could be a lot for the milking herd, one for dry cows and one for heifers post weaning.

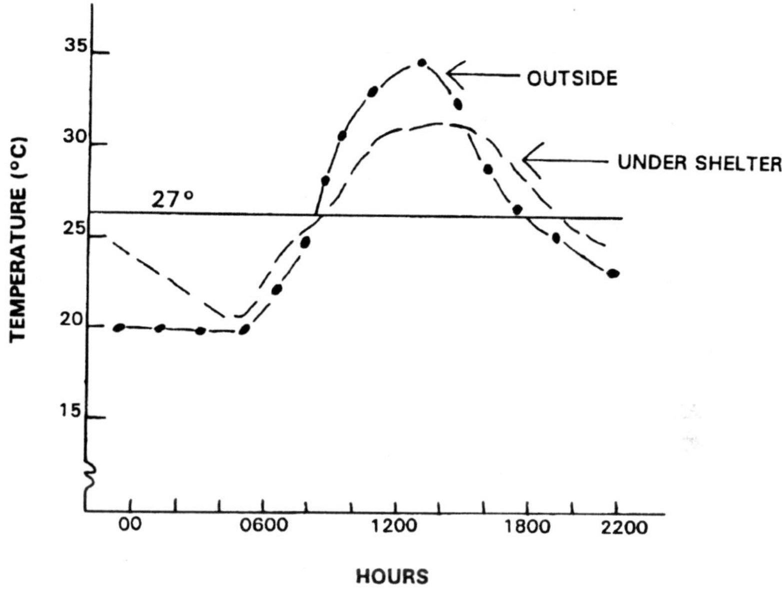

Figure 8.4. Air temperature for 24 hours outside and underneath shelter (McDowell, 1972).

In Figure 8.5, feed is offered along the fence line. There are those who argue that the feeding area should be covered. The author recommends no cover to discourage eating during mid-day as feeding can add to the heat load; feeding is best during early AM and late PM with the heaviest feeding of forages in the PM (Figure 2.4). The water is located away from the shelter to help maintain dryness under the shelter and to encourage cows to move periodically and consider relocating at the feed manager or outside if the temperature has declined.

Usually when we build barns or shelters, the roof is "A" shaped (closed at the top) or there may be an opening with a cap over top to permit hot air to rise out of the shelter. A roof structure used for a long while in Israel and now increasing in creditability for hot climates in general is depicted by the sketch in Figure 8.6. The roof slope is greater than usual (3 to 4° per 2.3 m) is recommended. The opening at the top should be around 50 cm.

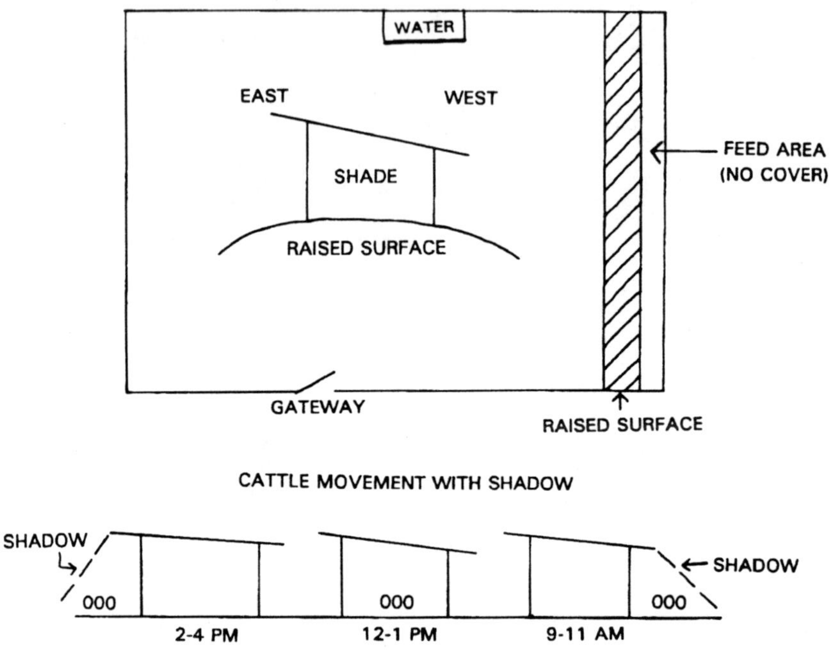

Figure 8.5. Lot organization to minimize effect of sun through use of shadow and radiation of body heat to cool side of atmosphere (000 = location of cows at various periods).

The steeper roof slope increases air flow across and above the roof. This creates negative pressure over the opening which then hastens the flow of air out the top and turbulence of air movement of air around the cows.

Cooling with use of fans and sprinklers is often recommended. Choice of research results from studies conducted thus far can favor their use or find no value over cost. In the feeding and comfort stall barn, fans appear a worthwhile investment. Orientation in mounting of the fans is important. Fans are most effective when located on the side of prevailing winds and "blow with the wind". When the barn is large, two or more sequences of fans may be needed. The fans should be aligned to promote

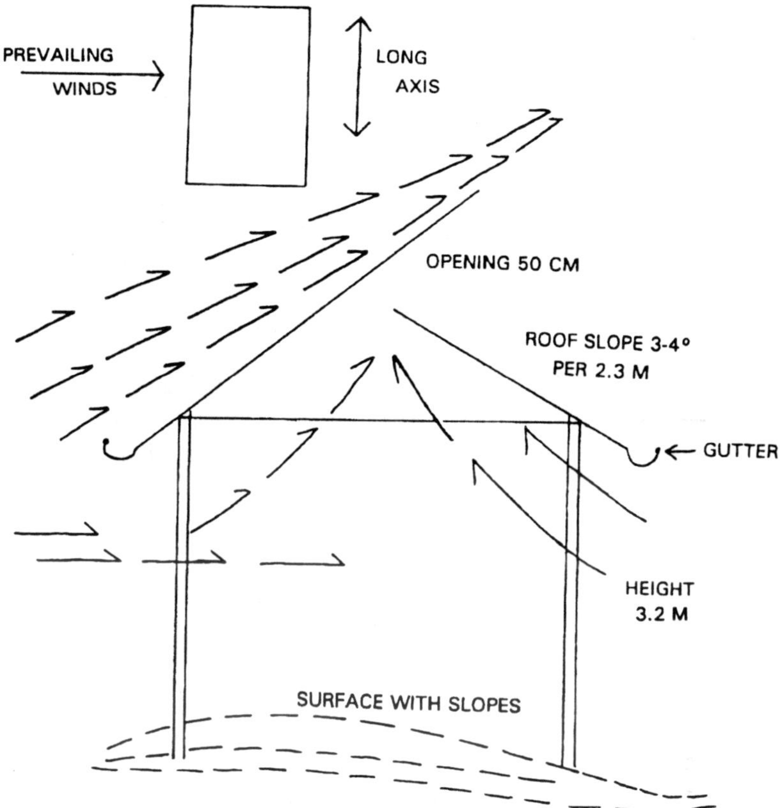

Figure 8.6. **Shelter or barn for "dynamic cooling" through orientation and roof slope to draw air underneath out the top at rapid rate due to low pressure above gap in roof.**

one direction movement, not have fans blowing "opposed to each other."

There are herd managers who strongly support the use of sprinklers for their cows; however, most of the research does not show significant gains in feed intake, milk yield or expression of estrus over a 24 hour period, in other words, the effects are "short term." Sprinklers have the disadvantages of increasing humidity level, increasing the wetness of floor surface which can attract cows to more readily lie down in messy areas. However, use in lower humidity areas of WCZ have generally been beneficial. Coupled with this is a high rate of disfunction unless the water line is filtered to remove particles which can clog the small openings of the sprayers.

Cows can be wetted and at the same time be cleaned by hosing with water in the holding area while awaiting entry into the milking parlor. If executed, it should be done far enough in advance for the cows to drip dry before milking otherwise as pointed out in Chapter VII the risk of mastitis will rise.

In brief, over investing in facilities is common in most of the WCZ. Lower cost is needed but most of all more consideration to cow comfort is required if she is expected to perform effectively.

Herd Records

Countries with a long history of dairying and those now developing viable commercial dairy enterprises attribute progress to herd, regional and national programs of recording, such as the DHI (Dairy Herd Improvement Program) in the U.S. or facsimile there of. The exception is India which uses a monitoring system through their Village Milk Cooperative Program. A recording program is vital to herd operators. The gathering of statistics on the performance of cattle and diseases is important to national governments for developing policies and is necessary to researchers in identifying problems and setting guidelines for improving animal efficiency. If herd owners wish to have a pool of trained manpower, either for on-farm work or as extension personnel, including veterinarians, they need to invest in record keeping.

In the past, the cost of record keeping was high because of cost of processing, but advancement in computer technology, particularly small or desk top computers, makes even national programs much lower in cost. At the national level, Taiwan is an excellent example. Dairy records processed on a desk top computer have been instrumental in obtaining government support to establish much needed central laboratories for determining milk composition (fat, protein, total solids), somatic cell counts on individual cows and herd feed composition.

Where governments have not seen fit to lend support to a recording program, Holstein or Jersey breeder associations have been effective sponsors. Argentina, Chile, Colombia, Puerto Rico, and Zimbabwe represent working examples. Personnel trained to visit farms to collect recordings of daily milk, breedings, and calvings have proven excellent for extending technology recommendations to the dairy farms. For instance, in Taiwan supervisors are very effective in getting herd owners to understand the value of records in herd management thereby rapid progress.

A herd record program such as DHI in the U.S., will of course meet opposition. Most responses are:

- "Recording costs too much." There is cost but a worthy investment in the future profitability of the herd.
- "I do not have time." Time is required but the herd manager will be able to prioritize daily tasks, e.g., instead of watching the entire herd for heats, concentrate on cows which need to breed.
- "I don't understand records." It is easy to learn.
- "My cows are not good enough." Your concern is recognized but your cows will fool you, especially when feeding according to production.
- "I can tell what a cow produces by her looks." When most dairy operators start records, they often find they ranked their cows entirely wrong using cow appearance.

1. <u>Use of Records in Herd Management</u>

When the author is invited to countries to offer suggestions on dairy operations, the general procedure is to visit a number of dairies. Before departing from the farms, comments after very short observations are requested. There are no records except the memory of the operator. In such situations the following are used:

- proportion of the cows chewing their cud while being observed taking into consideration time since last feeding

- percentage of cows with unbalanced udders by age group to estimate severity of mastitis
- estimate percent dry cows
- observe condition of feet and ease of movement of the cows
- identify feeds and approximate amounts offered
- availability of dirt exercise and resting area
- cleanliness in resting area
- body condition and approximate age of heifers for breeding
- percentage of shaggy hair coats, extended rumens and presence of nose discharge in calves

Comments are offered on general performance of the herd along with obvious strengths and weaknesses in management to include quality and quantity of feeds. If there is evidence of considerable mastitis, a number of cows are lame, 30% of the cows are dry and young stock unthrifty, it is obvious that the margin of income is low. Under these circumstances, it is difficult to arouse enthusiasm of the owner as he immediately translates suggestions, such as culling some of the blemished cows, better quality feed and regrading the dirt exercise lot to reduce mud into added costs and delayed financial returns on an already low margin of returns. The main reason for the farmer's disappointments is that he and the viewer are not seeing the operation the same way. He feels he is doing the best with the resources he has. In a herd, such as described, the only effective way to communicate with the owner would be through the use of numerical values gathered from records since both can see plus and minus values which can be related to economic measures. The real wish would be to seek the opinion of the cows on their evaluation of record keeping (Box 8.11).

Let's now switch to consultation for a herd with records on individual cows for one year or more. Hopefully, the records are available before the scheduled visit to the farm in order to prepare a comparison of the herd performance in a number of traits for what appears as reasonable goals, both in milk yield and supportive traits (Table 8.3). The herd is below the desired level in all measures, thus an opportunity to interact on a number of issues for deriving priorities on application of inputs. The herd is using progeny tested sires from AI. On the assumption that the genetics in the herd is satisfactory, discussions center on feeding and management.

The initial suspect of problems in the herd is feeding. The manager advises his offering the lactating cows feeds identified as Ration #1 in Table 8.4 with the expectation the cows will average 20 kg of milk per day but the actual is 16 kg. The questions are: Can we do better and not markedly increase costs? Based on estimated requirements, the cows should have sufficient intakes of DM, TDN, CP and P but deficient in calcium for maintenance of 525 kg of body weight and 20 kg of milk.

Table 8.3. Desired performance versus averages for a Holstein herd being evaluated.

Trait	Desired	Herd Average
Milk Yield (kg)		
1st lactation	4500 - 5300	4270
Later lactations	5400 - 6300	4480
Fat percent	3.4 - 3.6	3.0
Age of 1st calving, mo	≤ 30	32
Calving interval, days	410 - 420	448
Cow entry rate, heifers, %	25	29
No. calvings per cow	3.0 - 3.2	2.6
Culling		
Sterility	8 - 10	10
Bad feet	1 - 2	12
Calving to 1st breeding, days	< 90	137
Conception to 1st service, %	45	41
Services per conception	2.1	2.3
Days dry	70 - 80	111
Mature body weight, kg	510 - 540	525

Adapted from McDowell, 1991b.

Obviously, the feeds are not being utilized effectively. As fed, the ration appears low in DM (Napier grass, about 18% DM, brewers grain, distillers grains and soya waste all 20 to 25% DM). The ration is adequate in total fiber but quite low in chewable fiber (NDF); hence low saliva secretion to maintain satisfactory pH in the rumen. It is pointed out that the low pH could be causing less than desired digestion and acidosis may be the cause of foot problems and high rate of culling. The low calcium to phosphorus ratio can be affecting both milk yield and reproduction rate. Use of some dry forage is discussed coupled with other changes. Ration #2 (Table 8.4) is proposed. Using this ration, feed cost will be lowered and fat percent raised above 3.0 which would lift the penalty imposed at the processing center for substandard fat content. The feeding recommendations are accepted enthusiastically.

Table 8.4. Requirement for 525 kg Holstein cows to produce 20 kg of milk with 3.4% fat; ration currently in use (#1) and alternative (#2) to improve performance

	kilogram			gram		
	Feed/d	DM	TDN	CP	Ca	P
Maintenance			3.8	375	21	15
Milk production			6.0	1640	59	36
Total			9.8	2015	80	51
Ration #1						
Chopped Napier grass	30	5.4	2.9	421	19	16
Brewers grains, wet	6	1.5	1.0	375	5	8
Distillers grains, wet	6	1.5	1.3	375	2	11
Soya waste, wet	6	1.5	1.3	450	6	5
Concentrate	8	6.8	4.8	1088	24	25
Total	56	16.5	11.3	2709	56	65
Difference from needs			+1.5	+694	-24	+14
Ration #2						
Chopped Napier grass	20	3.6	1.9	280	13	11
Pangola grass hay	3	2.6	1.2	143	10	5
Brewers grains, wet	6	1.5	1.0	375	5	8
Soya waste, wet	6	1.5	1.3	450	6	5
Concentrate	8	6.8	4.8	1088	24	25
Calcium suppl. (1.5% DM)						
Total	44	16.0	10.2	2330	82	54
Difference from needs			+.4	+321	+2	+3

Adapted from McDowell, 1991b.

2. Foot Problems

It is pointed out that attention to care of feet could yield high returns. Examples are cited with the result that the herd owner is ready to invest, including having an extension agent come to the farm for a demonstration on foot trimming.

3. Breeding Efficiency

Values in Table 8.3 are used to show that at present breeding efficiency, as determined by all measures (age at first calving, calving interval, calving to time of first breeding, rate of conception to first service and services per conception) were below the desired level. Possible effects of the imbalances in the ration are cited as potentially limiting, e.g., lameness or "sensitivity in the feet" could be affecting reproduction rate as suggested by values in Table 8.5.

Table 8.5. Effects of lameness on reproduction of lactating Holsteins.

Measures	Deviation from expected
Calving to 1st service (days)	+4
Calving to conception (days)	+14
% conception to 1st service (%)	-10
Services per conception (no.)	+0.4
Culling rate for sterility (%)	+45

It is suggested that from observations in other Holstein herds, that the condition of the dirt exercise lot should be improved and the herd be kept on the lot 1 to 2 hours in the morning and a similar period in late afternoon as cows on dirt have about 33% more mounts and duration of estrus is longer. However the drainage situation on the lot needs improvement. Initially, the owner balks but sees economic returns in 30 to 60 days so becomes ready to invest.

Breeding and calving records on individual cows and barn records of breeding heifers are used. The current status of pregnancy shows a distinct seasonal effect; calvings July to January are more than double the numbers February to June. The heifers show three low months but these are not contiguous. The owner attributes this to numbers ready to breed. It is suggested that breeding of some heifers could be delayed to compensate for the low months for cows. Since cows calving February to May are slowest to rebreed during the hottest months and semen costs are high, it is suggested that it might be wise to delay breedings and concentrate on breeding heifers in these months. This pattern is being followed in the southeastern U.S. and parts of the WCZ, e.g., Puerto Rico.

4. <u>Heifers</u>

The herd owner agrees that an average of 32 months for first parturition is later than desired for Holsteins. A review of calvings in the herd during the last two years shows 64% calved at 30 months or earlier, 8% 31 to 35 months and 28% 36 to 40 months. This shows the mean is influenced by the slow breeding group, 36 to 40 months. After viewing the heifers, it is decided that level of forage feeding is low and the heifers are becoming overly fat. The heifers are tethered in stalls with less than 2 hours per day in the dirt lot possibly affecting regularity in the estrous cycle and conception. Shifting the heifer feeding to a program similar to that for the cows and placing the heifers in the lot during mid-day while the lactating herd is inside for shading is recommended and accepted.

Some heifers that calved recently are showing vaginal discharge. Interaction with the owner led to the conclusion that only 30% calved without assistance; 50% were given some assistance and for 20% a vet was brought in. Most heifers weigh 430 to 460 kg which is low for Holsteins to avoid calving difficulties. It was agreed that the proposed new program for feeding heifers could have high payout in better conditioning for parturition.

Approximately 10% of the young heifers are showing enlarged quarters indicating being suckled during calfhood. Fastening up for one hour after liquid feeding and offering some dry feed after fluid feeding is recommended and accepted. Two heifer calves appeared unthrifty. The owner recalled treating one for serious scouring and the other for pneumonia. It is suggested these be culled as they will in all probability be poor performers, if they reach lactating age. The sum on the heifers is that the owner now sees them more as the "future" of his enterprise and although not income producers, their care is extremely important.

5. <u>Mastitis</u>

The herd records included somatic cell count scores on individual cows. There were 27% of the cows scoring 5 to 9 for two months or longer. Projections of lactation yields to 305 days showed these cows are 10 to 28% below the herd average. The owner is advised that the high cell count cows are eating as much feed as these scoring lower, hence poor feed efficiency and higher cost per unit of milk sold.

The afternoon milking was observed. It was clear that the milkers were not drying the udder before applying the teat cups thereby permitting dirty water to be drawn into the teat cups during milking, a threat in the spread of bacteria. The potential for reduction of mastitis by treatment at the time of drying off was reviewed. The reduction of manure accumulation in the dirt surfaced exercise lot and adding fresh dirt and mounds to increase water run-off as part of a mastitis control program were discussed. For the first time, the herd owner saw procedures he could implement at no cost or low cost to increase returns.

In a general discussion following the review, it became clear that to obtain returns for use of high technology in genetic potential of the cows for milk yield, application of supporting technology to achieve economic returns on capital invested in genetics needed greater attention. Instead of feeling threatened from consultation, the owner felt helped in seeing problems both for the herd in general and individual cows. He saw readily how he could set priorities in use of labor and investment of capital to induce changes.

Record keeping in dairy enterprises is advised for all herds but especially in the WCZ as their value is higher. This is largely because there are as yet many unknowns on how best to manage dairy breeds for good economic and biological efficiency.

BIBLIOGRAPHY

Abubakar, B.Y. 1983. Procedures for estimating genetic progress for milk production in the tropics. M.S. thesis, Cornell University, Ithaca, NY.

Abubakar, B.Y., R.E. McDowell and L.D. Van Vleck. 1987a. Interaction of genotype and environment for breeding efficiency and milk production of Holsteins in Mexico and Colombia. Trop. Agric. (Trinidad) 64:17-22.

Abubakar, B.Y., R.E. McDowell, L.D. Van Vleck and E. Cabello. 1987b. Phenotypic and genetic parameters for Holsteins in Mexico. Trop. Agric. (Trinidad) 64:23-26.

Albright, J.L. 1993. Feeding behavior of dairy cattle. J. Dairy Sci. 76:485-498.

Barnett, A.J.G. 1954. Silage fermentation. New York Academic Press, Inc., New York.

Bauman, D.E. 1990. Bovine somatotropin: Review of emerging animal technology. Emerging agricultural technology: Issues for the 1990's. Office of Technology Assessment, U.S. Congress, Washington, DC.

Bauman, D.E. 1992. Bovine somatotropin: Review of an emerging animal technology. J. Dairy Sci. 75:3432-3451.

Bauman, D.E., S.N. McCutchen, W.D. Steinhour, P.J. Eppard and S. J. Sechen. 1985. Sources of variation and prospects for improvement of productive efficiency in the dairy cow: A review: J. Dairy Sci. 60:583-592.

Blake, R.W., F.J. Holmann, J. Gutierrez and G.F. Ce Vallov. 1988. Comparative profitability of U.S. Holstein artificial insemination sires in Mexico. Proc. Workshop: Dairying for profitability in Mexico, Mexico Holstein-Friesian Assoc., Queretaro.

Blake, R.W., M.A. Tomaszewski and C.R. Shumway. 1991. Profitable genetics: PV $ lists bulls for low risk and low cost. Dairy Herd Management 28(10):26.

Britt, J.H., R.G. Scott, J.D. Armstrong and M.D. Whitacre. 1986. Determination of estrous behavior in lactating Holstein cows. J. Dairy Sci. 69:2195.

Camoens, J.K., R.E. McDowell, L.D. Van Vleck and J.D. Rivera-Anaya. 1976. Holsteins in Puerto Rico. III. Components of variance associated with production traits and estimates of heritability. J. Agric. Univ. Puerto Rico 50:551-558.

Correa, M.T., H. Erb and J. Scarleff. 1993. Pathway analysis for seven postpartum disorders of Holstein cows. J. Dairy Sci. 76:1305-1312.

Ferreira, L. 1991. Managing for reproductive efficiency in the tropics. Proc. Int'l Conf.: Dairying in a tropical environment (IV). Holstein Assoc., U.S., Brattleboro, VT (p 135-142).

Flatt, W.P., L.A. Moore, N.W. Hooven and R.D. Plowman. 1965. Energy metabolism studies with a high producing lactating cow. J. Dairy Sci. 48:797-804.

Galina, C.S. and G. H. Arthur. 1991. Review of cattle reproduction in the tropics: Part 6: The male. Anim. Br. Abstr. 59:1-10.

Hahn, M.V., B.T. McDaniel and J.C. Wilk. 1986. Rates of hoof growth and wear in Holstein cattle. J. Dairy Sci. 69:2148-2156.

Hall, H.T.B. 1977. Diseases and parasites of livestock in the tropics. Longman, New York.

Henderson, C.R. 1971. Sire evaluation and its importance. Proc. Dairy Breeding Conf. Animal Science Mimeo. Series No. 13, Cornell University, Ithaca, New York.

Hoard's Dairyman. 1990. Herd health, W. D. Hoard & Sons Co., Fort Atkinson, Wisconsin.

Hoard's Dairyman. 1993. Control coliform mastitis with clean environment. Hoard's Dairyman 138(15) 636.

Hoffman, K., L.D. Muller, S.L. Fales and L.A. Holen. 1993. Quality evaluation and concentrate supplementation of rotational pasture grazed by lactating cows. J. Dairy Sci. 76:2651-2763.

Holman, F., R.W. Blake, R.A. Milligan, R. Barker, P.A. Oltnacu and M.V. Hahn. 1990a. Economic returns from United States artificial insemination sires in Holstein herds in Colombia, Mexico, and Venezuela. J. Dairy Sci. 73:2179-2189.

Holmann, F., R.W. Blake, M.V. Han, R. Barker, R.A. Milligan, P.A. Oltenacu and T.L. Stanton. 1990b. Comparative profitability of purebred and crossbred Holstein herds in Venezuela. J. Dairy Sci. 73:2190-2205.

Ladue, R.C. 1993. Vaccination and control programs. Proc. 2d Biennial Northeast Heifer Mgt. Symp. (p.53-58). Animal Sci. Mimeo No. 165, Cornell Univ., Ithaca, N.Y.

Lean, J.J., H.F. Troutt, M.L. Bruss, T.B. Farven, R.L. Baldwin, J.C. Galland, D. Kratzer, C.A. Holmbeng and L.D. Weaver. 1991. Postpartient metabolic and production responses in cows previously exposed in long-term treatment with somatotropin. J. Dairy Sci. 74:3429.

Lemka, L., R.E. McDowell, L.D. Van Vleck, H.A. Guha and J.J. Salazar. 1973. Reproductive efficiency and viability in two *Bos indicus* and two *Bos taurus* breeds in the tropics of India and Colombia. J. Anim. Sci. 36:644-652.

Mahana, W.C. 1990. How do these feeding thumb rules fit your herd? Hoard's Dairyman 135:30-31.

Makuza, S.M. 1993. Environmental and genetic factors affecting performance of Holsteins, Jerseys and Crossbreds in Zimbabwe. (unpublished data).

McCraw, R.L., K.R. Butcher, and B.T. McDaniel. 1980. Progeny tested sires compared with pedigree selected young sires. J. Dairy Sci. 63:1342.

McCullough, M.E. 1975. New trends in ensiling forages. World Animal Rev., No. 15, FAO, Rome.

McCullough, M.E. 1992. When we feed cows, we feed the rumen. Hoard's Dairyman 137:655.

McDaniel, B.T. 1981. Economic impact of calving difficulty in Holstein heifers. J. Dairy Sci. 65: Suppl. No. 1.

McDaniel, B.T. and W.E. Bell. 1992. Experimental comparison of systems of selecting Holstein bulls. Dept. of Animal Science Annual Rpt. 1992. Raleigh, NC.

McDaniel, B.T. and M.R. Dentine. 1985. Genetic gains in milk yield through artificial insemination and embryo transfer. Anais 1° Simpósio Internacional De Producão Animal, Sociedady Brasileira de Genética, Ribeirão Preto, Brazil.

McDaniel, B.T., B. Verbeek, J.C. Wilk, R.W. Everett and J.F. Keown. 1984. Relationships between hoof measures, stayabilities, reproduction and changes in milk yield from first to later lactation. J. Dairy Sci. 68: Suppl. No. 1.

McDowell, L.R., J.H. Conrad and F.G. Hembry. 1993. Minerals for grazing ruminants in tropical regions. Bull., 2nd ed. Animal Sci. Dept. Univ. Florida, Gainesville.

McDowell, R.E. 1972. Improvement of livestock production in warm climates. W. H. Freeman & Co., San Francisco, CA.

McDowell, R.E. 1982. Crossbreeding as a system of mating for dairy production. Southern cooperative Series Bul. No. 259, Louisiana State Univ., Baton Rouge.

McDowell, R.E. 1985. Meeting constraints to intensive dairying in tropical areas. Cornell Int'l Agri. Mimeo. No. 108, Cornell University, Ithaca, NY (41 p).

McDowell, R.E. 1987. Environmental and genetic factors influencing performance in Holsteins in warm climates. Proc. Workshop: Dairying in a tropical environment, Taichung, Taiwan, April 28-May 1, 1987 (pp 19-47). Holstein Assoc., Brattleboro, VT.

McDowell, R.E. 1988. An assessment of Jersey cattle: A review of research. Special study report to American Jersey Cattle Club, Columbus, Ohio.

McDowell, R.E. 1989. Meeting constraints to dairying in tropical areas. Proc. Int'l Seminar: Dairying in tropical environments, Padjadjaran Univ., Bandung, Indonesia, May 22-25, 1989 (pp 1-32). Holstein Assoc., Brattleboro, VT.

McDowell, R.E. 1991a. A partnership for humans and animals. Kinnic Publishers, Raleigh, NC.

McDowell, R.E. 1991b. Value of dairy records. Proc. Int'l Conf.: Dairying in a tropical environment, Tung Hai Univ., Taichung, Taiwan May 27-30, 1991 (pp 63-77). Holstein Assoc. Brattleboro, VT.

McDowell, R.E. 1992. Dairy production in developing countries. Proc. New Zealand Soc. Anim. Prod. 52:3-5.

McDowell, R.E. and K.B. Leining. 1978. Handling of European cattle in the tropics for reproduction efficiency. Proc. 2nd Int'l Congr. of the Chianina Breed. Sao Pulo, Brazil.

McDowell, R.E., G.R. Wiggans, J.K. Cameons, L.D. Van Vleck and D.G. St. louis. 1976. Sire comparisons for Holsteins in Mexico versus the United States and Canada. J. Dairy Sci. 59:298-304.

McDowell, R.E. et al. 1975. Tropical grass pastures with and without supplement for lactating cows in Puerto Rico. Univ., Puerto Rico Agri. Exp. Sta. Bul. 238, Rio Piedras, PR.

McDowell, R.E., N.W. Hooven and J.K. Cameons. 1976. Effect of climate on performance of Holsteins in first lactation. J. Dairy Sci.59:965-973.

Mechor, G.D. 1993. Treatment of diarrhea in the calf. Second Biennial Northeast Heifer Mgt. Symp. (p. 24-27). Animal Sci. Memeo Series No. 165, Cornell Univ., Ithaca, N.Y.

Mpofu, N., C. Smith, W. VanVuuren and E.b. Burnside. 1993. Breeding strategies for genetic improvement of dairy cattle in Zimbabwe. 2. Economic evaluation. J. Dairy Sci. 76:1173-1181.

Nat'l. Acad. Sci. 1981. Effects of environment on nutrient requirements of domestic animals. National Academy Press, Washington, DC.

Nebel, R.L. and M.L. McGillard. 1993. Interactions of high milk yield and reproductive performance in dairy cows. J. Dairy Sci. 76:3257-3268.

Nocek, J.E. 1993. Hoof care for dairy cattle. W.D. Hoard and Sons Co., Milwaukee Ave. West, Fort Atkinson, WI.

NRC. 1988. Nutrient requirements of dairy cattle, 6th Edition National Academy Press, Washington, DC.

Osei, S.A., K. Effah-Baah, and P. Karibari. 1991. The productive performance of Friesian cattle bred in the hot humid forest zone of Ghana. World Anim. Rev. 3:52.

Philpot, W.N.   1991.   Milking procedures and management practices for improving milk quality.   Proc. Int'l Conf. on Dairying in a Tropical Environment, Tung Hai Univ. Taichung, Taiwan. Holstein Assoc., Brattleboro, VT.

Philpot, W.N. and S.C. Nickerson. 1991. Mastitis: Counter attack, a strategy to combat mastitis.   Babson Bros. Co., Naperville, IL.

Politiek, R.D., O. Distl, T. Fjeldaas, J. Errres, B.T. McDaniel, E. Nielsen, D.J. Peteose, A. Reurink and P. Strandberg. 1986. Importance of claw quality in cattle: Review and recommendation to achieve genetic improvement. Livestock Prod. Sci. 15:133-152.

Powell, R.L. 1992. Bull evaluations for Holstein bulls used in Mexico, USDA-DHIA Mexican Bull Evaluations, USDA, Beltsville, Maryland.

Romero, L.C.   1986.   Environmental and genetic factors influencing the performance of Holsteins in Puerto Rico. M.S. thesis, Cornell University, Ithaca, NY.

Romero, L.C., T.J. Trebilcock, Jr., R.W. Everett and R.E. McDowell. 1989. Evaluation of Holstein sires used in DHIA herds of Puerto Rico. J. Agric. Univ. Puerto Rico 73:406-417.

Reed, J.D., B.S. Cooper and P. J.H. Neate. 1988. Plant breeding and nutritive value of crop residues.   Proc. Workshop, ILCA, Addis Ababa, Ethiopia.

Ruppel, K.A.   1993.   Quality water: The most important nutrient. Hoard's Dairyman 138:94-95.

Schneeberger, C.P. 1982. Factors affecting the performances of a Holstein herd in Maracay, Venezuela. PhD Thesis. Cornell Univ. Ithaca, N.Y.

Senger, P. 1990. These steps will improve your heat detection. Hoard's Dairyman 135:646-647.

Shook, G.E. 1989. Selection for disease resistance. J. Dairy Sci. 72:1349-1362.

Silva, H.M., C.J. Wilcox, W.W. Thatcher, R.B. Becker and D. Morse. 1992. Factors affecting days open, gestation length and calving interval in Florida dairy cattle. J. Dairy Sci. 75:288-293.

Spalding, R.W., R.W. Everett and R.H. Foote. 1975. Fertility in New York artificially inseminated Holstein herds in Dairy Herd Improvement. J. Dairy Sci. 58:718-723.

Seykora, A.J., and B.T. McDaniel. 1983. Heritabilities and correlations of lactation yields and fertility for Holsteins. J. Dairy Sci. 66:1486.

Sitorus, P. 1982. Productivity of Friesian and their offspring resulting from outbreeding with imported semen. Proc. 2nd World Genetics Congr.

Stanton, T.L., R.W. Blake, R.L. Quaas and L.D. Van Vleck. 1991. Response to selection of United States Holstein sires in Latin America. J. Dairy Sci. 74:651-664.

Stevenson, J. 1993. Here are tips for improving heat detection. Hoard's Dairyman 138:101.

Talbott, C.W., R.E. McDowell, B.T. McDaniel and M. Zaman. 1992. Comparison of dairy merit from three Bos taurus breeds crossed with Pakistani Sahiwals. J. Dairy Sci. 75:Suppl.1, p.155.

Vailes, L.D., and J.H. Britt. 1990. Influence of footing surface on mounting and other sexual behaviors of estrual Holstein cows. J. Animal Sci. 68:914-921.

Van Soest, P.J. 1982. Nutritional ecology of the ruminant. O & B Books, Inc. 1215 NW Kline Place, Corvallis, Oregon.

Van Vleck, L.D. 1977. Theoretical and actual genetic progress in dairy cattle. Proc. Int'l Conf. on Quantitative Genetics, Iowa State Univ., Ames, Iowa.

Van Vleck, L.D. 1981. Potential genetic impact of artifical insemination, sex selection, embryo transfer, cloning and selfing in dairy cattle. New technologies in animal breeding, Academic Press, 1991.

Vinson, W.E. 1983. Selection for non-yield traits of economic importance in dairy cattle. Holstein Science Rpt., Holstein Assoc., Brattleboro, VT.

Weaver, L.D., J. Gall and U. Sasnik and P. Cowen. 1986. Factors affecting embryo transfer success in recipient heifers under field conditions. J. Dairy Sci. 69:2711.

Wiggans, G.R. and R.E. McDowell. 1993. Statistics are databanks in animal science. In: The literature of animal science and health, W.C. Olsen ed., Cornell University Press, Ithaca, (p. 52-69).

Yakstis, J.J. 1983. Parasites: Evaluating the problem. Anim. Nutrition and Health 39:24-33.

Yazman, J.A. 1980. Feeding systems for lactating cows and growing heifers in a tropical environment. Ph.D. Thesis, Cornell University, Ithaca, NY.

Yazman, J.A., R.E. McDowell, H. Cestero, J.A. Arroyo-Aguili, J.D. Rivera-Anaya, M. Soldevila and F. Roman-Garcia. 1982., Efficiency of utilization of tropical grass pastures by lactating cows with and without supplement. J. of Agri. Univ. Puerto Rico 66:200-222.

3 1896 00100 3866